## Advanced Lectures in Mathematics

For further volumes: http://www.springer.com/series/12121

Wolfgang Ebeling

# Lattices and Codes

## A Course Partially Based on Lectures by Friedrich Hirzebruch

3rd Edition

Wolfgang Ebeling
Burgdorf, Germany
ebeling@math.uni-hannover.de

ISBN 978-3-658-00359-3 ISBN 978-3-658-00360-9 (eBook)
DOI 10.1007/978-3-658-00360-9

Library of Congress Control Number: 2012947439

Mathematics Subject Classification:
11H06, 11H31, 11H55, 11F11, 11F41, 11R04, 11R18, 94B05, 94B15, 94B75, 51F15, 51E10

Springer Spektrum


*Editorial Office:* Ulrike Schmickler-Hirzebruch | Barbara Gerlach

Printed on acid-free paper.

Springer Spektrum is a brand of Springer DE.
Springer DE is part of Fachverlagsgruppe Springer Science+BusinessMedia.
www.springer-spektrum.de

*To the memory of Friedrich Hirzebruch*
*with great admiration*

# Preface to the Third Edition

It is pleasing that this book seems to be still sufficiently popular to warrant a third edition and has proven to be suitable for master courses. I recently gave such a course and while preparing for it I found that there were still some mistakes in the book and that the notation was not always consistent. Moreover, recently a question which was stated as an open problem in the book has been solved [67]. Therefore I have taken the opportunity to make the necessary changes and to improve and update the book.

I would like to thank Elisabeth Werner for pointing out that the codes in Sect. 2.10 which were erroneously called Reed-Solomon codes in the previous editions are in fact Hamming codes. This has now been corrected. I am grateful to Richard Borcherds for suggesting that I include a very short proof of the uniqueness of the Leech lattice as an application of the results in Chapter 4. This has been added as Remark 4.2. Finally the layout of the book has been changed.

I would like to thank Springer Spektrum, especially Ulrike Schmickler-Hirzebruch, for making a new edition of the book possible.

Sadly, on May 27 this year, Friedrich Hirzebruch, on whose lectures this book is partially based, passed away. I would like to express my gratitude and my admiration by dedicating this book to his memory.

Hannover, July 2012 *Wolfgang Ebeling*

# Preface to the Second Edition

The changes from the first edition are as follows. Numerous corrections and improvements have been made. More basic material is included to make the text even more self-contained and to keep the prerequisites to a minimum. Automorphism groups of lattices and codes are treated in more detail. In particular, a new section, Sect. 4.5, on Conway's result about a relation between the Leech lattice and the automorphism group of the 26-dimensional unimodular hyperbolic lattice has been added. There are some hints to new results. Finally, several new exercises have been added.

A short course on some of the main topics presented in detail in the book can be found in [23].

I would like to thank R. E. Borcherds, G. van der Geer, F. Heckenbach, B. Herzog, H. Koch, J. H. van Lint, V. Remmert, P. Schenzel, J. Spandaw, Z.-X. Wan, W. Willems, and J. Zintl for sending me or communicating to me corrections, comments, and suggestions for improvements. In particular, I am grateful to J. Spandaw for pointing out an error in the proof of Proposition 1.2 and for his suggestion of a correct proof which I followed.

Last but not least I would like to thank Ulrike Schmickler-Hirzebruch from Vieweg for her support and encouragement during the preparation of the second (and also of the first) edition of the book.

Hannover, May 2002 *Wolfgang Ebeling*

# Preface to the First Edition

The purpose of coding theory is the design of efficient systems for the transmission of information. The mathematical treatment leads to certain finite structures: the error-correcting codes. Surprisingly problems which are interesting for the design of codes turn out to be closely related to problems studied partly earlier and independently in pure mathematics. This book is about an example of such a connection: the relation between codes and lattices. Lattices are studied in number theory and in the geometry of numbers. Many problems about codes have their counterpart in problems about lattices and sphere packings. We give a detailed introduction to these relations including recent results of G. van der Geer and F. Hirzebruch.

Let us explain the history of this book. In [58] J. S. Leon, V. Pless, and N. J. A. Sloane considered the Lee weight enumerators of self-dual codes over the prime field of characteristic 5. They wrote in the introduction to their paper: "The weight enumerator of any one of the codes ... is strongly constrained: it must be invariant under a three-dimensional representation of the icosahedral group. These invariants were already known to Felix Klein, and the consequences for coding theory were discovered by Gleason and Pierce (and independently by the third author) ... (It is worth mentioning that precisely the same invariants have recently been studied by Hirzebruch in connection with cusps of the Hilbert modular surface associated with $\mathbb{Q}(\sqrt{5})$. However, there does not appear to be any connection between this work and ours.)" In 1985 G. van der Geer and F. Hirzebruch held a summer school on coding theory in Alpbach in the Austrian alps. There they found such a connection.

In the winter semester 1986/87 Hirzebruch gave a course at the University of Bonn entitled "Kodierungstheorie und Beziehungen zur Geometrie" ("Coding theory and relations with geometry"). Among other things he explained the above mentioned results. Some lectures were given by N. J. A. Sloane and N.-P. Skoruppa. In January 1987 the author gave his inaugural lecture on "The Leech lattice".

When the author came to Eindhoven in October 1988 he gave a course entitled "Lattices and codes". He lectured two hours a week from October 1988 until April 1989. This course was partially based on notes of Hirzebruch's lectures taken by Thomas Höfer and Constantin Kahn and on an unpublished note by Nils-Peter Skoruppa [83]. The aim of this course was to discuss the relations between lattices and codes and to provide all the necessary prerequisites and examples to be able to understand the result of van der Geer and Hirzebruch. During the course notes were prepared and distributed. These formed the basis for the present book.

A lattice $\Gamma$ in Euclidean $n$-space $\mathbb{R}^n$ is a discrete subgroup of $\mathbb{R}^n$ with compact quotient $\mathbb{R}^n/\Gamma$. The relation between lattices and binary codes has been studied by a number of authors (cf. e.g. [60], [84], [85], and [4]). It is one of the themes of the book of J. H. Conway and N. J. A. Sloane [21]. There is some overlap with that book, but whereas [21] is mainly a collection of original articles, we have tried to write a coherent and detailed introduction to this topic. Of course we had to make a choice, and there is much more material in [21] which we don't mention. The main topics we have chosen for are the connections between weight enumerators of codes and theta functions of lattices, the classification of even unimodular 24-dimensional lattices, and the Leech lattice. The results of van der Geer and Hirzebruch can be considered as a generalization of the results on weight enumerators of codes and theta functions of lattices in the binary case. They are based on a relation between lattices over integers of number fields and $p$-ary codes, $p$ being an odd prime number. The only reference for these results is a letter of Hirzebruch to Sloane reprinted in [36]. We give a detailed account on these results.

Another example of a relation between coding theory and a seemingly unrelated branch of pure mathematics is the link between coding theory and algebraic geometry recently established by Goppa's construction of codes using algebraic curves. This was also mentioned in Hirzebruch's lectures but it is not treated in this book. We refer the interested reader to [57].

We now give a survey of the contents of this book. The book has five chapters.

In the first chapter we give the basic definitions for a lattice and for a code. We explain the fundamental construction of a lattice $\Gamma_C$ in $\mathbb{R}^n$ from a binary linear code $C$. This leads to a correspondence between (doubly even) self-dual binary linear codes of length $n$ and certain (even) unimodular lattices in $\mathbb{R}^n$. The basic example is the extended binary Hamming code of length 8, which corresponds to the famous lattice of type $E_8$. The $E_8$-lattice is an example of a root lattice, i.e., a lattice generated by vectors $x \in \mathbb{R}^n$ with $x^2 = 2$. The root lattices play an important rôle in our book. They are studied and classified in Sect. 1.4. We determine which root lattices come from codes. Some additional facts about root lattices which will be needed in Chapter 4 are collected in Sect. 1.5.

Chapter 2 deals with important functions associated to lattices and codes. On the one hand one has the theta function of a lattice. From the theta function one can compute the number of lattice points on spheres around the origin. We show that the theta function of an even unimodular lattice in $\mathbb{R}^n$ is a modular form of weight $\frac{n}{2}$. For the proof of this fact we have included introductory sections on modular forms. In the presentation of this material we are influenced by Serre's book [81]. On the other hand one has the (Hamming) weight enumerator polynomial of a code. We discuss several applications of the theory of modular forms to weight enumerators of codes via the correspondence between lattices and codes. First we derive a certain relation between the weight enumerator coefficients of a doubly even self-dual code of length 24. We are lead to the following question: do there exist doubly even self-dual codes of length 24 which have no codewords of weight 4? In Sect. 2.8 we show the uniqueness and existence of such a code: it is the extended binary Golay code (our proof follows [7]). Analogously we ask for the existence of an even unimodular lattice in $\mathbb{R}^{24}$ which contains no roots, i.e., vectors $x$ with $x^2 = 2$. Using the Golay code we construct such a lattice: the Leech lattice. In Sect. 2.9 we present a relation between the Hamming weight enumerator of a binary linear code $C$ and the theta function of the corresponding lattice $\Gamma_C$. We deduce the MacWilliams identity and Gleason's theorem. Finally we discuss extremal codes and extremal lattices. In Sect. 2.10 we consider an important class of codes, the quadratic residue codes. Some of the corresponding extended codes are extremal. We recover again the Hamming code and the Golay code.

In Chapter 3 we present Venkov's proof [89] that Niemeier's enumeration [68] of the even unimodular 24-dimensional lattices is correct. For this purpose we first study modified theta functions, namely theta functions with spherical coefficients, and their behaviour under transformations of the modular group. These are results due to E. Hecke [32] and B. Schoeneberg ([78], [79]). They are used to classify root sublattices of even unimodular lattices. In dimension 24 we derive that the root sublattice of an even unimodular lattice in $\mathbb{R}^{24}$ is one of a list of 24 root lattices. Each root lattice of this list can be realized in one and only one way up to isomorphism. In this way we get Niemeier's classification of the even unimodular 24-dimensional lattices. We also prove that the correspondence $C \mapsto \Gamma_C$ induces a one-to-one correspondence between equivalence classes of doubly even codes $C$ in $\mathbb{F}_2^n$ and isomorphism classes of even lattices in $\mathbb{R}^n$ containing a root lattice of type $nA_1$. This was observed by H. Koch [46]. Now 9 of the 24 Niemeier lattices contain such a root lattice, and hence they correspond to binary codes.

One of the 24 even unimodular 24-dimensional lattices is the Leech lattice, which does not contain any roots at all. Chapter 4 is devoted to this lattice. We first show the uniqueness of this lattice. Then we discuss the sphere packing and covering determined by the Leech lattice. J. Leech conjectured soon after the discovery of this lattice that the covering radius of this lattice is equal to $\sqrt{2}$ times its packing radius. This was proved by J. H. Conway, R. A. Parker, and N. J. A. Sloane in 1982 [17]. Their method of proof involved finding all the deep holes in the Leech lattice, i.e., all points in $\mathbb{R}^{24}$ that have maximal distance from the lattice. They discovered that there are precisely 23 distinct types of deep holes, and that they are in one-to-one correspondence with the 23 even unimodular lattices in $\mathbb{R}^{24}$ containing roots. The original proof (reprinted in [21]) partly consists of rather long case-by-case verifications. Later R. E. Borcherds [5] gave a uniform proof by embedding the Leech lattice in a hyperbolic lattice of rank 26. We indicate Borcherds' proof in Sect. 4.4. We also mention (in Sect. 4.3) the 23 constructions for the Leech lattice, one for each of the deep holes or Niemeier lattices, found by Conway and Sloane [20].

So far we have only considered binary codes. In Chapter 5 we present a generalization of the results of Chapter 2 to self-dual codes over the prime field $\mathbb{F}_p$, where $p$ is an odd prime number. In the binary case we constructed an integral lattice in $\mathbb{R}^n$ from a binary linear code of length $n$. We generalize this construction by associating a lattice over the integers of a cyclotomic field to a code over $\mathbb{F}_p$. The necessary basic facts from algebraic number theory are provided and lattices over integers of cyclotomic fields are studied in Sect. 5.1. The construction is given in Sect. 5.2. In Sect. 5.3 we consider theta functions for such lattices as studied e.g. by H. D. Kloosterman [43] and M. Eichler [24]. These theta functions depend on $\frac{p-1}{2}$ variables $z_l, l = 1, \dots, \frac{p-1}{2}$. They are Hilbert modular forms. We state and prove the theorem of van der Geer and Hirzebruch which gives a connection between the Lee weight enumerator of a self-dual linear code $C$ over $\mathbb{F}_p$ and the theta function of its associated lattice $\Gamma_C$. In Sect. 5.4 – Sect. 5.6 we give applications of this result. In the case of ternary codes ($p = 3$) a well-known result of M. Broué and M. Enguehard [4] is derived. We consider the equation of the tetrahedron and, with a short digression to the (excluded) case $p = 4$, the equation of the cube. Finally we turn to the case $p = 5$ and explain the relation between the paper [34] of Hirzebruch and the paper [58] by Leon, Pless and Sloane mentioned in the beginning. We conclude this chapter by indicating in Sect. 5.7 how one can verify that certain theta functions are Hilbert modular forms. The Sect. 5.7 is essentially a translation into English of the unpublished paper [83] of N.-P. Skoruppa. Parts of this paper have also been used in Sect. 3.1.

The book is partially based on, but not identical with Hirzebruch's lectures, as it is also expressed in the different title. The following topics of these lectures are not included in this book: the determination of the number of doubly even self-dual codes of length $n$ and, as already mentioned, a section on Riemann-Roch for curves and applications in coding theory. On the other hand, we have added Chapter 4 on the Leech lattice. Moreover, our presentation is in many places more detailed and contains among other things in addition: a proof of the existence of a fundamental system of roots for any root lattice (Sect. 1.4), and the proof of the uniqueness of the binary Golay code and Conway's construction of it (Sect. 2.8).

We have tried to keep the prerequisites as minimal as possible. The original course was intended for third year students having the German 'Vordiplom'. A basic knowledge in algebra and complex analysis is required. Several exercises can be found in the text.

Theorems, Propositions, Lemmas, etc. are numbered consecutively within each chapter. So e.g. Theorem 1.3 refers to the third theorem in Chapter 1. Each chapter is divided into sections. The 4-th section in Chapter 5 is referred to as Sect. 5.4.

The author would like to thank Dr. Thomas Höfer and Dr. Constantin Kahn for making their notes of Hirzebruch's course available to him.

He is grateful to the participants of his course in Eindhoven for their interest and criticism. He would like to thank Prof. Dr. A. E. Brouwer, Prof. Dr. J. H. van Lint, and Prof. Dr. J. J. Seidel for many useful hints. For example he learned from Prof. van Lint the introduction of the Golay code as included in this book. He would like to express his thanks to all his former colleagues at Eindhoven for their support.

The author is grateful to Prof. Dr. F. Hirzebruch, Dr. Ruth Kellerhals, and Dr. Niels-Peter Skoruppa for reading parts of an older version of the manuscript, for their criticism and helpful remarks. He is particularly indebted to Prof. Dr. F. Hirzebruch for permission to use notes of his course and for his help and encouragement during the preparation of this book.

The author would like to thank Kai-Martin Knaak and Robert Wetke for setting the manuscript into TEX.

Hannover, January 1994 *Wolfgang Ebeling*

# Contents

# Chapter 1
# Lattices and Codes

## 1.1 Lattices

In this section we introduce the basic concept of a lattice in $\mathbb{R}^n$. For references see [81], [61], [9], [45], and [72].

**Definition.** A *lattice* in $\mathbb{R}^n$ is a subset $\Gamma \subset \mathbb{R}^n$ with the property that there exists a basis $(e_1,\ldots,e_n)$ of $\mathbb{R}^n$ such that $\Gamma = \mathbb{Z}e_1 \oplus \ldots \oplus \mathbb{Z}e_n$, i.e., $\Gamma$ consists of all integral linear combinations of the vectors $e_1,\ldots,e_n$.

**Example 1.1** The standard example is the lattice

$$\mathbb{Z}^n \subset \mathbb{R}^n$$

consisting of all points in $\mathbb{R}^n$ with integral coordinates. Here $(e_1,\ldots,e_n)$ is the standard basis of $\mathbb{R}^n$ (see Fig. 1.1).

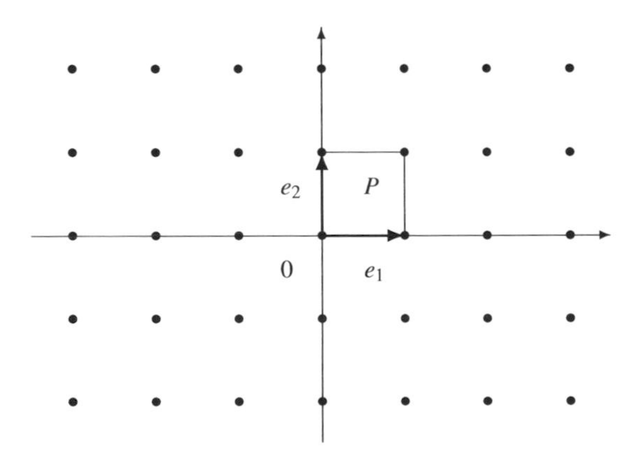


**Fig. 1.1** Standard lattice $\mathbb{Z}^2 \subset \mathbb{R}^2$

Let $\Gamma$ be a lattice in $\mathbb{R}^n$. A basis $(e_1,\ldots,e_n)$ of $\mathbb{R}^n$ with $\Gamma = \mathbb{Z}e_1 \oplus \ldots \oplus \mathbb{Z}e_n$ is called a *basis* of $\Gamma$. The quotient $\mathbb{R}^n/\Gamma$ is an $n$-dimensional torus. It is obtained by identifying the faces of the *fundamental parallelotope*

$$P = \{\lambda_1 e_1 + \ldots + \lambda_n e_n \mid 0 \le \lambda_i \le 1\}.$$

In particular, the quotient $\mathbb{R}^n/\Gamma$ is compact. Conversely, by a theorem of Bieberbach, a discrete subgroup $\Gamma \subset \mathbb{R}^n$ with compact quotient $\mathbb{R}^n/\Gamma$ is a lattice [80].

The *volume* of a lattice is

$$\text{vol}\,(\mathbb{R}^n/\Gamma) = \text{vol}\,(P) = |\det((e_1,\ldots,e_n))|$$

where $((e_1,\ldots,e_n))$ is the matrix whose rows are the vectors $e_1,\ldots,e_n$. This is true for the standard lattice $\mathbb{Z}^n \subset \mathbb{R}^n$, where $P$ is the $n$-dimensional unit cube. It follows in general from the transformation formula of integrals, since $((e_1,\ldots,e_n))$ is the transformation matrix of $\mathbb{Z}^n$ into $\Gamma$.

More generally, let $\Gamma' \subset \mathbb{R}^n$ be a lattice with $\Gamma' \subset \Gamma$. Then clearly the index $|\Gamma/\Gamma'|$ is finite and

$$\text{vol}\,\left(\mathbb{R}^n/\Gamma'\right) = \text{vol}\,(\mathbb{R}^n/\Gamma)\left|\Gamma/\Gamma'\right|.$$

We denote the Euclidean scalar product of two vectors $x,y \in \mathbb{R}^n$ by $x \cdot y$. So

$$x \cdot y = \sum_{i=1}^{n} x_i y_i\,.$$

The Euclidean scalar product is a non-degenerate, positive definite, symmetric bilinear form. Put $a_{ij} = e_i \cdot e_j$ and let $A$ be the matrix $((a_{ij}))$. Let $C$ be the matrix $((e_1,\ldots,e_n))$. Then $A = CC^t$. Therefore

$$\text{vol}\,(P) = |\det C| = \sqrt{\det A}\,.$$

Let $V = \mathbb{R}^n$. We identify $V$ with the dual vector space $V^* = \text{Hom}\,(V,\mathbb{R})$ by means of the mapping $V \longrightarrow V^*$, $x \longmapsto f_x$, with $f_x(y) = x \cdot y$. Let $\Gamma$ be a lattice in $\mathbb{R}^n$. We denote the *dual lattice* of $\Gamma$ by $\Gamma^*$. It is

$$\begin{aligned}\Gamma^* &= \text{Hom}\,(\Gamma,\mathbb{Z})\\ &= \{x \in \mathbb{R}^n \mid x \cdot y \in \mathbb{Z} \text{ for all } y \in \Gamma\}.\end{aligned}$$

Let $(e_1,\ldots,e_n)$ be a basis of $\Gamma$, and let $(e_1^*,\ldots,e_n^*)$ be the dual basis, i.e., $e_i^* \cdot e_j = \delta_{ij}$. Then

$$e_i^* = \sum_{j=1}^{n} b_{ij} e_j$$

and $B = ((b_{ij})) = A^{-1}$. The $e_i^*$ form a basis of $\Gamma^*$. One easily computes that the matrix $((e_i^* \cdot e_j^*))$ is equal to $B$. This shows in particular that

$$\text{vol}\,(\mathbb{R}^n/\Gamma^*) = \frac{1}{\text{vol}\,(\mathbb{R}^n/\Gamma)}\,.$$

A lattice $\Gamma \subset \mathbb{R}^n$ is called *integral*, if $x \cdot y \in \mathbb{Z}$ for all $x, y \in \Gamma$. Another characterization of integral lattices is the following.

**Definition.** Let $R$ be a commutative ring with unity. A *symmetric bilinear form module* $(S,b)$ *over* $R$ is a pair consisting of a free $R$-module $S$ of rank $n$ (i.e., isomorphic to $R^n$) and of a symmetric bilinear form $b : S \times S \longrightarrow R$.

**Proposition 1.1.** *The integral lattices in* $\mathbb{R}^n$ *are precisely the symmetric bilinear form modules* $(S,b)$ *over the ring of integers* $\mathbb{Z}$*, where* $b : S \times S \longrightarrow \mathbb{Z}$ *is a positive definite symmetric bilinear form.*

*Proof.* Let $\Gamma \subset \mathbb{R}^n$ be an integral lattice. Then $\Gamma$ is a free $\mathbb{Z}$-module of rank $n$ and the Euclidean scalar product is a positive definite symmetric bilinear form on $\Gamma$ with values in $\mathbb{Z}$.

On the other hand, let $(S,b)$ be a symmetric bilinear form module over $\mathbb{Z}$ with a positive definite symmetric bilinear form $b$. Consider the real vector space $V := S \otimes \mathbb{R}$. Then $b$ extends to a scalar product on $V$. Let $(\varepsilon_1, \ldots, \varepsilon_n)$ be an orthonormal basis of $V$ with respect to the scalar product $b$, and identify $V$ with $\mathbb{R}^n$ by mapping $(\varepsilon_1, \ldots, \varepsilon_n)$ to the standard basis of $\mathbb{R}^n$. Then $b$ becomes the Euclidean scalar product of $\mathbb{R}^n$ and $S \subset \mathbb{R}^n$ is an integral lattice. This proves Proposition 1.1. □

Two symmetric bilinear form modules $(S,b)$ and $(S',b')$ over $R$ are called *isomorphic* if there is an $R$-linear bijection $h : S \to S'$ satisfying $b'(h(x),h(y)) = b(x,y)$ for all $x,y \in S$. Let $(S,b)$ be a symmetric bilinear form module over $R$. An $R$-linear bijection $h : S \to S$ satisfying $b(h(x),h(y)) = b(x,y)$ for all $x,y \in S$ is called an *automorphism* of $(S,b)$. The automorphisms of $(S,b)$ form a group $\mathrm{Aut}(S,b)$ called the *automorphism group* of $(S,b)$. By Proposition 1.1, these notions carry over to integral lattices in $\mathbb{R}^n$.

Also without the assumption that $b$ is positive definite, a symmetric bilinear form module $(S,b)$ over $\mathbb{Z}$ is called an *integral lattice*. In the first three chapters of this book, lattices will be lattices in $\mathbb{R}^n$. But in Chapter 4 we shall also consider some indefinite integral lattices. Moreover, in Chapter 5 we shall also study symmetric bilinear form modules (lattices) over the ring of integers of a certain algebraic number field.

Now let $\Gamma$ be an integral lattice with basis $(e_1, \ldots, e_n)$, and let $A$ be the matrix $A = ((e_i \cdot e_j))$. Then $A$ is an integral matrix and the determinant $\det A$ of $A$ is an integer. This integer is independent of the choice of the basis. For let $(\widetilde{e}_1, \ldots, \widetilde{e}_n)$ be another basis of $\Gamma$, and let $\widetilde{A} = ((\widetilde{e}_i \cdot \widetilde{e}_j))$. Then

$$\widetilde{e}_i = \sum_{j=1}^{n} q_{ij} e_j ,$$

and the matrix $Q = ((q_{ij}))$ is in $\mathrm{GL}_n(\mathbb{Z})$, which means in particular that $\det Q = \pm 1$. But

$$\widetilde{A} = QAQ^t ,$$

so $\det \widetilde{A} = \det A$. So $\det A$ does not depend on the choice of the basis $(e_1, \ldots, e_n)$, and this number is also called the *discriminant* of the lattice $\Gamma$, written $\mathrm{disc}\,(\Gamma)$. We have already seen that

$$\text{vol}\,(\mathbb{R}^n/\Gamma) = \sqrt{\text{disc}\,(\Gamma)},$$
$$\text{vol}\,(\mathbb{R}^n/\Gamma^*) = \frac{1}{\sqrt{\text{disc}\,(\Gamma)}}.$$

From the formula

$$\text{vol}\,(\mathbb{R}^n/\Gamma) = \text{vol}\,(\mathbb{R}^n/\Gamma^*)\cdot|\Gamma^*/\Gamma|$$

we derive the following formula for the discriminant of $\Gamma$:

$$\text{disc}\,(\Gamma) = |\Gamma^*/\Gamma|\,.$$

A lattice $\Gamma \subset \mathbb{R}^n$ is called *unimodular* if $\Gamma^* = \Gamma$. This amounts to saying that $\Gamma$ is an integral lattice, and that the canonical homomorphism $V \longrightarrow V^*$, $x \longmapsto f_x$ with $f_x(y) = x\cdot y$, maps $\Gamma$ onto $\Gamma^*$.

Analogously, let $(S,b)$ be a symmetric bilinear form module over a ring $R$. The *dual module* $S^*$ of $S$ is the module $\text{Hom}_R(S,R)$. The symmetric bilinear form module $(S,b)$ is called *unimodular*, if the canonical homomorphism $S \longrightarrow S^*$, $x \longmapsto b(x,\ )$, is bijective.

In matrix terms, this means the following. Let $\Gamma$ be an integral lattice with basis $(e_1,\ldots,e_n)$. Then $\Gamma$ is unimodular if and only if the matrix $A = ((e_i\cdot e_j))$ has integer coefficients and its determinant disc $(\Gamma)$ is equal to 1. The last condition is equivalent to

$$\text{vol}\,(P) = 1\,.$$

Let $\Gamma$ be an integral lattice in $\mathbb{R}^n$. A $\mathbb{Z}$-submodule $\Lambda$ of $\Gamma$ is called a *sublattice* of $\Gamma$. It is a lattice in some subspace $W \subset \mathbb{R}^n$ which is isomorphic to $\mathbb{R}^k$ for some $k$. In particular, the dual lattice $\Lambda^*$ is defined to be

$$\Lambda^* = \{x \in W \mid x\cdot y \in \mathbb{Z} \text{ for all } y \in \Lambda\}.$$

A sublattice $\Lambda$ of $\Gamma$ is called *primitive* if $\Gamma/\Lambda$ is a free $\mathbb{Z}$-module. If $K$ is a subset of $\Gamma$ we call the $\mathbb{Z}$-submodule

$$K^\perp = \{y \in \Gamma \mid x\cdot y = 0 \text{ for all } x \in K\}$$

the sublattice *orthogonal* to $K$.

Let $\Lambda_1,\ldots,\Lambda_m$ be sublattices of $\Gamma$. The lattice $\Gamma$ is called the *orthogonal direct sum* of the sublattices $\Lambda_1,\ldots,\Lambda_m$, denoted by $\Gamma = \Lambda_1 \perp \ldots \perp \Lambda_m$, if $\Gamma$ is the direct sum of the submodules $\Lambda_1,\ldots,\Lambda_m$ and $x\cdot y = 0$ for all $x \in \Lambda_i$, $y \in \Lambda_j$, and $i \neq j$.

**Proposition 1.2.** *Let $\Gamma$ be a unimodular lattice in $\mathbb{R}^n$ and $\Lambda$ be a primitive sublattice of $\Gamma$. Then*

$$\text{disc}\,(\Lambda) = \text{disc}\,(\Lambda^\perp).$$

*Proof.* (Cf. also [59, (2.3)].) We show that there is a natural group isomorphism $\Gamma/(\Lambda \perp \Lambda^\perp) \to \Lambda^*/\Lambda$. Since $\Gamma/\Lambda$ is torsion free, the mapping $i^* : \Gamma^* \to \Lambda^*$ dual to the inclusion $i : \Lambda \to \Gamma$ is surjective. Since $\Gamma$ is unimodular, we have $\Gamma^* = \Gamma$ and we get a surjective mapping $i^* : \Gamma \to \Lambda^*$. Factoring both sides by $\Lambda$, the mapping $i^*$ induces a surjective mapping

$$\iota : \Gamma/\Lambda \to \Lambda^*/\Lambda.$$

We claim that the kernel of this mapping is the image $I$ of $\Lambda^\perp$ in $\Gamma/\Lambda$. Since $i^*(\Lambda^\perp) = 0$, it is clear that $I \subset \ker \iota$. Conversely, suppose that $x \in \Gamma$ represents an element of the kernel of $\iota$. Then there exists a $y \in \Lambda$ such that $i^*(x-y) = 0$. But this implies $x - y \in \Lambda^\perp$. Hence the class of $x$ in $\Gamma/\Lambda$ lies in $I$. Therefore we get an isomorphism

$$\Gamma/(\Lambda \perp \Lambda^\perp) \to \Lambda^*/\Lambda.$$

Since $\Lambda$ is primitive, the same is true for $\Lambda^\perp$, as one can easily see. Moreover, $(\Lambda^\perp)^\perp = \Lambda$. Therefore we can interchange the rôles of $\Lambda$ and $\Lambda^\perp$ and find a group isomorphism $\Gamma/(\Lambda \perp \Lambda^\perp) \to (\Lambda^\perp)^*/\Lambda^\perp$. In particular

$$\mathrm{disc}\,(\Lambda^\perp) = |(\Lambda^\perp)^*/\Lambda^\perp| = |\Gamma/(\Lambda \perp \Lambda^\perp)| = |\Lambda^*/\Lambda| = \mathrm{disc}\,(\Lambda),$$

what had to be shown. □

**Exercise 1.1** Let $\Gamma$ be a lattice in $\mathbb{R}^n$ and $\Lambda \subset \Gamma$ be a unimodular sublattice. Show that $\Gamma = \Lambda \perp \Lambda^\perp$.

An integral lattice $\Gamma$ is called *even* if $x^2 = x \cdot x \equiv 0 \pmod 2$ for all $x \in \Gamma$. In matrix terms, this means that the diagonal elements $e_i \cdot e_i$ of the matrix $A$ are all even. Analogously, a symmetric bilinear form module $(S, b)$ over $\mathbb{Z}$ is called *even* if $b(x,x) \equiv 0 \pmod 2$ for all $x \in S$.

## 1.2 Codes

In this section we define the concept of a code. As a general reference for coding theory we refer to the book of J.H. van Lint [56]. We partly follow [37].

Error correcting codes are important for the transmission of information, e.g. in telephone and satellite communication and in compact disc players. The problem is that in any kind of communication the original message might not be received correctly because of noise or transmission errors. Therefore the message is encoded with a certain redundancy such that in the process of decoding possible transmission errors can be detected and corrected. The design of such error correcting codes is the subject of coding theory.

Mathematically a message is a finite sequence of symbols from some alphabet, for which we take a finite field $\mathbb{F}_q$ with $q = p^r$ ($p$ prime) elements. In communication and computer technology one frequently uses for obvious reasons the field $\mathbb{F}_2$, i.e., the letters $0, 1$. Each of these letters is called a "bit". Encoding can be considered as an injective mapping $f : \mathbb{F}_q^k \to \mathbb{F}_q^n$ where $n > k > 0$. The image of the mapping $f$, $f(\mathbb{F}_q^k) =: C \subset \mathbb{F}_q^n$, is called a *code* (of length $n$).

**Definition.** A *code $C$ of length $n$* is a nonempty proper subset of $\mathbb{F}_q^n$.

If $|C| = 1$ the code is called *trivial*. Trivial codes are excluded in the sequel. If $q = 2$ the code is called a *binary* code, for $q = 3$ a *ternary* code, etc. The elements of $C$ are called *codewords*, and $n$ is called the *wordlength* of $C$.

In a simplified picture, with the use of codes the transmission of information consists of three essential steps as follows:

| message | | | | | | | decoded message |
|---|---|---|---|---|---|---|---|
| $\to$ | encoder | $\to$ | channel | $\to$ | decoder | | $\to$ |
| $a \in \mathbb{F}_q^k$ | | $f(a) \in C$ | | $b$ | | | $g(b) \in C$ |

The message is put into a coding device called the encoder. The message can be data sampled in certain periods from some acoustic signals, the degree of blackness of some square of a grid for the recognition of a picture (e.g. a satellite picture) or part of data to be stored in a computer memory. The message leaves the encoder as a codeword $f(a)$ of a code $C$; technically this may be a sequence of impulses of different amplitude. Such a codeword now passes through an information channel with noise. An information channel can be an electromagnetic field, but it can also stand for the process of saving and subsequent loading of a binary word in a computer. The word noise should also not be taken literally; it can mean e.g. electrical interference, dust or scratches on a compact disc or the bombardment of a computer chip by $\alpha$-particles. As an effect of this noise the codeword arriving at the decoder at the receiver may be falsified. Of course it would be desirable to detect possible errors. To enable this, the idea is to choose the code $C$ in such a way that the codewords differ from each other as much as possible. Then the decoder looks for the codeword which most resembles the received word.

A very simple example is the binary repetition code of length 3. Here $q = 2$, $k = 1$, $n = 3$, $C = \{(0,0,0),(1,1,1)\} \subset \mathbb{F}_2^3$. The information to be transmitted is "yes" or "no". "Yes" is encoded in $(1,1,1)$, "no" in $(0,0,0)$. If the decoder receives $(0,1,0)$, then it is reasonable to assume that "no" was transmitted, because $(0,1,0)$ differs only at one position from $(0,0,0)$ but at two positions from $(1,1,1)$. Hence the output of the decoder will be "no". One easily sees that this code corrects *one* error, i.e., if there is a transmission error at at most one position, then this will be detected and corrected in any case.

In order to describe this phenomenon in a quantitative manner, we introduce the following notions. Let $x = (x_1,\ldots,x_n) \in \mathbb{F}_q^n$. The *weight* $w(x)$ of $x$ is the number of nonzero $x_i$. If $x \in \mathbb{F}_q^n$, $y \in \mathbb{F}_q^n$, then the *(Hamming) distance* $d(x,y)$ of $x$ and $y$ is defined by $d(x,y) := w(x-y)$.

Let $C$ be a nontrivial code. The minimum of the distances $d(x,y)$ for $x,y \in C$, $x \neq y$, is called the *minimum distance* of the code $C$. An $(n,M,d)$-code is a code with wordlength $n$, $M$ codewords, and minimum distance $d$. E.g. the repetition code of length 3 is a $(3,2,3)$-code.

One can easily see that a code with minimum distance $d$ can correct exactly $t$ errors where $t$ is defined by $d = 2t+1$ for $d$ odd, resp. by $d = 2(t+1)$ for $d$ even.

To ensure a correct transmission of information one is, therefore, interested in codes with a big minimum distance $d$. The repetition code has $d = 3$, but it has a disadvantage. It is too wasteful: 3 bits for the transmission of a one-bit-information. A spacecraft is subject to energy and time limits; computer memory should not be wasted. This leads to the fundamental problem of coding theory: In order to achieve large minimum distance one needs a big alphabet or words with many positions, in any case a large expenditure. But this is again undesirable.

In mathematical terms one introduces besides the minimum distance $d$ also the *information rate* $R$ of a code in $\mathbb{F}_q^n$:

$$R = \frac{\log_q |C|}{\log_q |\mathbb{F}_q^n|}.$$

Here $|C|$ resp. $|\mathbb{F}_q^n|$ denotes the number of codewords resp. of all possible words. If $C$ is a binary code, the number $\log_2 |C|$ can be interpreted as the minimal number of yes-no-decisions which one needs to locate each codeword. The repetition code of length 3 has information rate $\frac{1}{3}$.

With these notions the main problem of coding theory can be "formulated" as follows: Find codes where the two incompatible objectives of a big minimum distance $d$ and a big information rate $R$ are realized.

Presumably the historically first of the more subtle and efficient codes is the *Hamming code* ([29], [30]). (A nice reference for the history of this code is [88].) This is a binary code of length 7. It is defined by the mapping

$$\begin{array}{ccc} \mathbb{F}_2^4 & \to & \mathbb{F}_2^7 \\ (x_1,x_2,x_3,x_4) & \mapsto & (x_1,x_2,x_3,x_4,x_5,x_6,x_7) \end{array}$$

where $x_5$, $x_6$, and $x_7$ are determined in such a way that the sum of the 4 elements in a circle of Fig. 1.2 is always 0. This yields 3 linearly independent linear equations in $\mathbb{F}_2^7$.

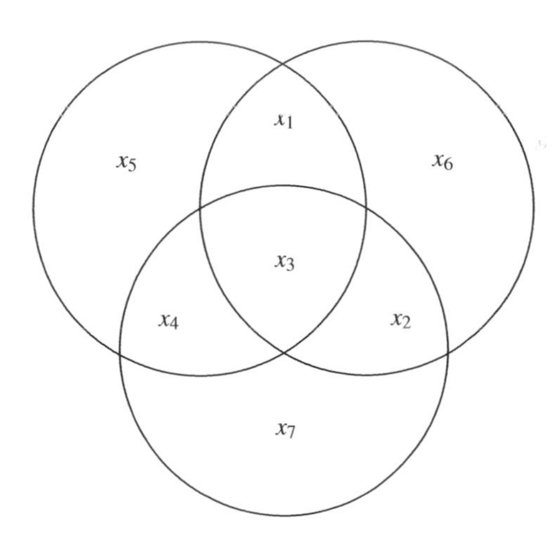


**Fig. 1.2** The conditions defining the Hamming code: the sum of the elements in a circle has to be equal to zero

The set of solutions is called $H$. This is a 4-dimensional linear subspace of $\mathbb{F}_2^7$. Therefore the code is a so called linear code.

**Definition.** A *linear code* $C$ is a linear subspace of $\mathbb{F}_q^n$.

If $C$ is a linear code, $k$ is the dimension of $C$ and $d$ is its minimum distance, then $C$ is called an $[n,k,d]$-code.

For a linear code $C$ the minimum distance is equal to the minimum weight, i.e., to the minimum of the weights of nonzero codewords.

Consider the vectors of $\mathbb{F}_q^n$ as column vectors. Then a linear code is defined by an exact sequence

$$0 \to \mathbb{F}_q^k \xrightarrow{A} \mathbb{F}_q^n \xrightarrow{B} \mathbb{F}_q^{n-k} \to 0$$

where $A$ and $B$ are linear mappings. The exactness of the sequence is equivalent to the three conditions rank $A = k$, $BA = 0$, rank $B = n-k$. The code $C$ defined by this sequence can be obtained in two ways.

First $C = A(\mathbb{F}_q^k) \subset \mathbb{F}_q^n$. The linear mapping $A$ is given by an $n \times k$ matrix $A$. The *columns* of $A$ form a basis of $C$. Usually one considers the transpose $G = A^t$ of $A$; this $k \times n$ matrix for which the rows form a basis of $C$ is called a *generator matrix* of $C$.

On the other hand $C = \ker B$, i.e., $x \in C$ if and only if $Bx = 0$. The linear mapping $B$ is given by an $(n-k) \times n$ matrix $B$. The *rows* of $B$ are the relations defining $C$. The matrix $B$ is called a *parity check matrix* of $C$. For every $x \in \mathbb{F}_q^n$ we call $Bx \in \mathbb{F}_q^{n-k}$ the *syndrome* of $x$. The codewords of $C$ are characterized by syndrome 0. We have defined the Hamming code by indicating the syndromes. In this case, if at most one error can occur and if an error occurs, then the syndrome indicates the position at which the error occurred.

Let $C$ be a linear code defined by an exact sequence as above. From linear algebra we know that a linear mapping $\phi : V \to W$ between vector spaces $V$ and $W$ induces a dual mapping $\phi^* : W^* \to V^*$ between the corresponding dual spaces $W^*$ and $V^*$; if $V$ and $W$ are finite dimensional, then we can identify the vector spaces with their corresponding dual spaces after the choice of bases. Therefore the above sequence induces a dual sequence

$$0 \to \mathbb{F}_q^{n-k} \xrightarrow{B^t} \mathbb{F}_q^n \xrightarrow{A^t} \mathbb{F}_q^k \to 0.$$

The condition $BA = 0$ is equivalent to the condition $A^t B^t = 0$. This exact sequence defines the *dual code* $C^\perp$, i.e., $C^\perp := B^t(\mathbb{F}_q^{n-k})$. If $C$ has dimension $k$ then $C^\perp$ has dimension $n-k$.

For $x, y \in \mathbb{F}_q^n$ we define their scalar product $x \cdot y$ by

$$x \cdot y := \sum x_i y_i.$$

**Lemma 1.1.** $C^\perp = \{y \in \mathbb{F}_q^n \mid x \cdot y = 0 \textit{ for all } x \in C\}$.

*Proof.* By definition, $y \in C^\perp$ if and only if $A^t y = 0$. But $A^t y = 0$ is equivalent to $A^t y \cdot z = 0$ for all $z \in \mathbb{F}_q^k$. Now $A^t y \cdot z = y^t A z = y \cdot Az$. Therefore $x \in C^\perp$ if and only if $x \cdot y = 0$ for all $y \in C = A(\mathbb{F}_q^k)$, which proves the lemma. □

A linear code $C$ is called *self-dual* if and only if $C = C^\perp$. Note that $\dim C + \dim C^\perp = n$, so $C = C^\perp$ implies that $n$ is even, $\dim C = \frac{n}{2}$ and $C \subset C^\perp$. This is also sufficient for a code $C$ to be self-dual. Note also that $C \subset C^\perp$ if and only if $x \cdot y = 0$ for all $x, y \in C$.

A binary code $C$ is called *doubly even*, if the weights $w(x)$ of all codewords $x \in C$ are divisible by 4. A doubly even code $C$ satisfies $C \subset C^\perp$, since over $\mathbb{Z}$

$$x \cdot y = \frac{1}{2}\left((x+y)^2 - x^2 - y^2\right).$$

The symmetric group $\mathscr{S}_n$ operates on $\mathbb{F}_q^n$ by permutation of the coordinate positions. Two codes $C$ and $C'$ over $\mathbb{F}_q^n$ are called *equivalent*, if they have both wordlength $n$ and if there is a $\sigma \in \mathscr{S}_n$ with $\sigma(C) = C'$. The *automorphism group* $\mathrm{Aut}(C)$ of a code $C$ is defined by

$$\mathrm{Aut}(C) := \{\sigma \in \mathscr{S}_n \mid \sigma(C) = C\}.$$

Now we return to the Hamming code $H$ of length 7. The $2^4 = 16$ codewords of the Hamming code are

0000000 1000110

0001101 1001011

0010111 1010001

0011010 1011100

0100011 1100101

0101110 1101000

0110100 1110010

0111001 1111111

One sees that the minimum weight is 3, therefore $H$ is a $[7,4,3]$-code. There is 1 word of weight 0, 7 words of weight 3, 7 words of weight 4, and one word of weight 7.

The information rate of the Hamming code is $R = \frac{4}{7}$. If one would use the repetition code with the same minimum distance as the Hamming code for the transmission of 4-bit-words, one would have to repeat each 4-bit-word three times. The information rate of this repetition code is $\frac{4}{12}$. The information rate of the Hamming code is therefore $\frac{12}{7} \approx 1.71$ times as big as the one of the repetition code and that means that the Hamming code can transmit 71% more information than the repetition code during the same time.

As a set of generators of $H$ one can take the 7 codewords of weight 3. These correspond to seven 3-element-subsets of a 7-element-set with the property that any two have exactly one element in common. (They form a so called 2-$(7,3,1)$-design, see Sect. 2.8.) They correspond to the lines of a projective plane of order 2

$$\mathbb{P}_2(\mathbb{F}_2) = \mathbb{F}_2^2 \cup \mathbb{F}_2 \cup \{\infty\}.$$

The seven points and seven lines of this plane form the configuration of Fig. 1.3. The incidence matrix is

$$\begin{pmatrix} 0&1&1&0&1&0&0 \\ 0&0&1&1&0&1&0 \\ 0&0&0&1&1&0&1 \\ 1&0&0&0&1&1&0 \\ 0&1&0&0&0&1&1 \\ 1&0&1&0&0&0&1 \\ 1&1&0&1&0&0&0 \end{pmatrix}$$

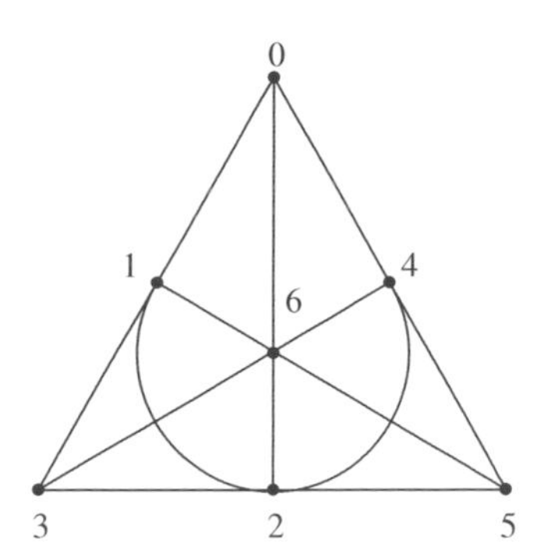


**Fig. 1.3** Configuration of the lines of $\mathbb{P}_2(\mathbb{F}_2)$

In this way we recover the seven codewords of weight 3 of $H$. Hence a submatrix of this matrix is a generator matrix for $H$. Note that the 1's in the first row are at the positions 1,2, and 4 which are the quadratic residues modulo 7 (see also Sect. 2.10). The other codewords are the 7 vectors obtained from the rows by interchanging 0's and 1's, the zero vector and the vector with all components equal to 1.

The automorphism group of the Hamming code is the automorphism group of the projective plane $\mathbb{P}_2(\mathbb{F}_2)$ and hence equal to the group $\mathrm{GL}_3(\mathbb{F}_2)$, which is a simple group of order 168 studied by F. Klein.

The Hamming code is not self-dual, but it can be extended to a self-dual code of length 8 by a general recipe. Let $C \subset \mathbb{F}_2^n$ be a binary code of length $n$. Consider the mapping

$$\begin{array}{cccc} I: & \mathbb{F}_2^n & \to & \mathbb{F}_2^{n+1} \\ & (x_1,\ldots,x_n) & \mapsto & (x_1,\ldots,x_n,x_1+\ldots+x_n) \end{array}.$$

The code $\widetilde{C} := I(C)$ is called the *extended code* . This means that we have added an extra overall parity check to the code $C$. The extended code $\widetilde{H} := I(H)$ of the Hamming code is called the *extended Hamming code* . This is an $[8,4,4]$-code with 1 codeword of weight 0, 14 of weight 4, and 1 of weight 8. Thus it is doubly even and self-dual, since all weights are divisible by 4.

## 1.3 From Codes to Lattices

From binary linear codes we can construct lattices. Take the standard lattice $\mathbb{Z}^n \subset \mathbb{R}^n$ and consider the reduction mod 2:

$$\rho : \mathbb{Z}^n \longrightarrow (\mathbb{Z}/2\mathbb{Z})^n = \mathbb{F}_2^n.$$

This is a group homomorphism. Let $C$ be an $[n,k,d]$-code. Since $\mathbb{F}_2^n/C \cong \mathbb{F}_2^{n-k}$, $C$ is a subgroup of index

$$|\mathbb{F}_2^n/C| = 2^{n-k}$$

of $\mathbb{F}_2^n$. Therefore $\rho^{-1}(C)$, the preimage of $C$ in $\mathbb{Z}^n$, is a subgroup of index $2^{n-k}$ of $\mathbb{Z}^n$. In particular $\rho^{-1}(C)$ is a free abelian group of rank $n$. Therefore $\rho^{-1}(C)$ is a lattice in $\mathbb{R}^n$. One has

$$\begin{aligned} \mathrm{vol}\,(\mathbb{R}^n/\rho^{-1}(C)) &= |\mathbb{Z}^n/\rho^{-1}(C)|\ \mathrm{vol}\,(\mathbb{R}^n/\mathbb{Z}^n) \\ &= 2^{n-k}. \end{aligned}$$

**Definition.** $\Gamma_C := \frac{1}{\sqrt{2}}\rho^{-1}(C)$.

The set $\Gamma_C$ is a lattice in $\mathbb{R}^n$. Let $x,y \in \Gamma_C$. Then $x$ and $y$ can be written

$$x = \frac{1}{\sqrt{2}}(c+2z),\ y = \frac{1}{\sqrt{2}}(c'+2z')$$

for some $c,c' \in \{0,1\}^n$ representing codewords in $C$ and some $z,z' \in \mathbb{Z}^n$. By abuse of notation we shall identify in the sequel $\mathbb{F}_2^n$ with the subset $\{0,1\}^n$ of $\mathbb{Z}^n$ and write briefly $c,c' \in C$. Now

$$x^2 = \frac{1}{2}(c^2 + 4cz + 4z^2)$$

and

$$\begin{aligned} x \cdot y &= \frac{1}{2}((x+y)^2 - x^2 - y^2) \\ &\equiv \frac{1}{4}((c+c')^2 - c^2 - (c')^2) \pmod{\mathbb{Z}} \\ &\equiv \frac{1}{2} c \cdot c' \pmod{\mathbb{Z}}. \end{aligned}$$

It follows that $x \cdot y \in \mathbb{Z}$ for all $x,y \in \Gamma_C$ if and only if $c \cdot c' \in 2\mathbb{Z}$ for all $c,c' \in C$. Therefore $\Gamma_C$ is an integral lattice if and only if $C \subset C^\perp$. Moreover, we see that $x^2 \in 2\mathbb{Z}$ for all $x \in \Gamma_C$ if and only if $c^2 \in 4\mathbb{Z}$ for all $c \in C$. This means that $\Gamma_C$ is even if and only if $C$ is doubly even.

Finally we see that

$$\begin{aligned} & k = \frac{n}{2} \\ \Longleftrightarrow\ & \mathrm{vol}\,(\mathbb{R}^n/\rho^{-1}(C)) = 2^{n/2} \\ \Longleftrightarrow\ & \mathrm{vol}\,(\mathbb{R}^n/\Gamma_C) = 1. \end{aligned}$$

From this it follows that $C$ is self-dual if and only if $\Gamma_C$ is a unimodular integral lattice. So we have shown:

**Proposition 1.3.** *Let $C$ be a linear code.*

(i) *$C \subset C^\perp$ if and only if $\Gamma_C$ is an integral lattice.*

(ii) *$C$ is doubly even if and only if $\Gamma_C$ is an even lattice.*
(iii) *$C$ is self-dual if and only if $\Gamma_C$ is unimodular.*

As an example consider the extended Hamming code $\widetilde{H}$. Then $\Gamma_{\widetilde{H}}$ is an even unimodular lattice in $\mathbb{R}^8$. We want to determine a basis of this lattice. First we consider the seven vectors determined by the rows of the matrix defining the Hamming code

$$f_1 := \frac{1}{\sqrt{2}}(0,1,1,0,1,0,0,1)$$
$$f_2 := \frac{1}{\sqrt{2}}(0,0,1,1,0,1,0,1)$$
$$f_3 := \frac{1}{\sqrt{2}}(0,0,0,1,1,0,1,1)$$
$$f_4 := \frac{1}{\sqrt{2}}(1,0,0,0,1,1,0,1)$$
$$f_5 := \frac{1}{\sqrt{2}}(0,1,0,0,0,1,1,1)$$
$$f_6 := \frac{1}{\sqrt{2}}(1,0,1,0,0,0,1,1)$$
$$f_7 := \frac{1}{\sqrt{2}}(1,1,0,1,0,0,0,1)$$

Then $f_i^2 = 2$ for all $1 \leq i \leq 7$. Moreover $f_i \cdot f_j = 1$ for all $1 \leq i, j \leq 7$ with $i \neq j$, since any two lines of $\mathbb{P}_2(\mathbb{F}_2)$ intersect in a point. Put

$$e_1 = f_1,\ e_2 = f_2 - f_1,\ e_3 = f_3 - f_2,\ e_4 = f_4 - f_3,$$
$$e_5 = f_5 - f_4,\ e_6 = f_6 - f_5,\ e_7 = f_7 - f_6.$$

Then $e_i^2 = 2$ for all $1 \leq i \leq 7$, and $e_i \cdot e_j \in \{0, -1\}$ for all $1 \leq i, j \leq 7$, $i \neq j$. In order to describe the matrix $((e_i \cdot e_j))$, we associate a graph to these elements as follows. We represent each vector $e_i$ by a vertex. We connect the vertices corresponding to $e_i$ and $e_j$ for $i \neq j$ by an edge if and only if $e_i \cdot e_j = -1$. This graph is called the *Coxeter-Dynkin diagram* corresponding to $\{e_1, \ldots, e_7\}$. It is the following graph

$$\overset{e_1}{\bullet} — \overset{e_2}{\bullet} — \overset{e_3}{\bullet} — \overset{e_4}{\bullet} — \overset{e_5}{\bullet} — \overset{e_6}{\bullet} — \overset{e_7}{\bullet}$$

It follows that the vectors $e_1, \ldots, e_7$ are linearly independent.

Next we consider the line through the points 2,3, and 5 in Fig. 1.3 as the line at infinity and take the complement of this line where we attach the following weights to the points in Fig. 1.3:

$$0 \mapsto -1$$
$$1 \mapsto -1$$
$$4 \mapsto 1$$
$$6 \mapsto -1\,.$$

This means that we take the vector

$$e_8 := \frac{1}{\sqrt{2}}(-1,-1,0,0,1,0,-1,0) \in \Gamma_{\widetilde{H}}.$$

Clearly $e_8^2 = 2$. It follows from the construction that

$$e_8 \cdot f_i = \begin{cases} 0 & \text{for } i = 1,2,3,4, \\ -1 & \text{for } i = 5,6,7. \end{cases}$$

Therefore the Coxeter-Dynkin diagram corresponding to $\{e_1, \dots, e_8\}$ is the graph

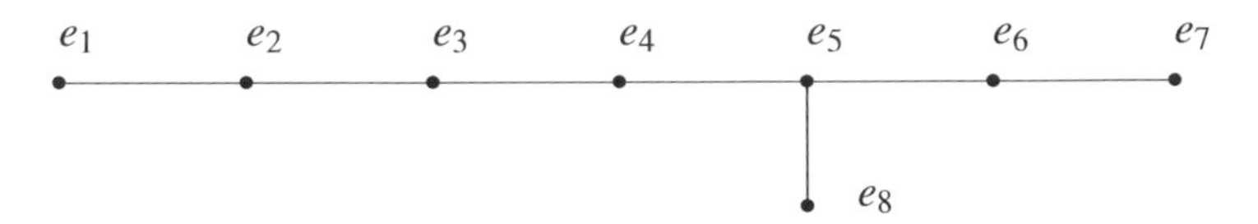


This graph is the Coxeter-Dynkin diagram of an exceptional Lie group of type $E_8$. In particular $\det((e_i \cdot e_j)) = 1$. Therefore $e_1, \dots, e_8$ form a basis of $\Gamma_{\widetilde{H}}$. We call the lattice $\Gamma_{\widetilde{H}}$ the *$E_8$-lattice*.

We determine the number of vectors $x \in \Gamma_{\widetilde{H}}$ with $x^2 = 2$. This is equal to the number of vectors $y \in \rho^{-1}(\widetilde{H})$ with $y^2 = 4$. Write $y = c + 2z$ with $c \in \widetilde{H}$ and $z \in \mathbb{Z}^n$ as above. There are 14 possibilities for $c$ with $c^2 = 4$. In each case one can replace any of the 1's by $-1$'s. Since each $c$ with $c^2 = 4$ contains exactly 4 ones, this yields 16 possibilities for each $c$. For $c = 0$ one has $y^2 = 4z^2$. There are $8 \cdot 2$ possibilities for $z \in \mathbb{Z}^n$ with $z^2 = 1$. Therefore there are

$$14 \cdot 16 + 8 \cdot 2 = 240$$

vectors $x \in \Gamma_{\widetilde{H}}$ with $x^2 = 2$.

We recall some facts from group theory. Let $G$ be a group with identity element 1, and let $X$ be a set. We say that $G$ *acts* on $X$ if there is a mapping $\phi : G \times X \to X$ such that

(A1) $\phi(g_1, \phi(g_2, x)) = \phi(g_1 g_2, x)$ for all $g_1, g_2 \in G$ and $x \in X$;
(A2) $\phi(1, x) = x$ for all $x \in X$.

These two conditions imply that $\phi$ induces a group homomorphism from $G$ to the symmetric group on $X$.

Let $G$ and $H$ be groups and assume that $G$ acts as a group of automorphisms on $H$. Let $\phi : G \times H \to H$ be the corresponding group action. The basic example is when $G$ and $H$ are subgroups of a larger group $G'$, $G$ normalizes $H$, and the action of $G$ on $H$ is by inner automorphisms: $\phi(g,h) = ghg^{-1}$. The *semi-direct product* $H \cdot G$ is defined as follows: as a set it is $H \times G$, and the multiplication is defined by

$$(h_1, g_1)(h_2, g_2) = (h_1 \phi(g_1, h_2), g_1 g_2).$$

**Exercise 1.2** Check that this makes $H \cdot G$ a group.

We have a short exact sequence of group homomorphisms

$$1 \to H \xrightarrow{i} H \cdot G \xrightarrow{p} G \to 1$$

where $i$ and $p$ are defined by

$$i(h) = (h, 1), \quad p(h, g) = g.$$

A short exact sequence of group homomorphisms

$$1 \to G' \xrightarrow{\alpha} G'' \xrightarrow{\beta} G''' \to 1$$

is called *split* if $\beta$ has a right inverse, i.e., if there exists a homomorphism $\beta' : G''' \to G''$ such that $\beta \circ \beta' = \mathrm{id}_{G'''}$. One also says that the sequence *splits*.

**Lemma 1.2.** *Let*

$$1 \to H \xrightarrow{\alpha} G' \xrightarrow{\beta} G \to 1$$

*be an exact sequence of group homomorphisms where G acts on H. Then this sequence splits if and only if G′ is isomorphic to the semi-direct product of H and G.*

*Proof.* First assume that this sequence splits. If $\beta' : G \to G'$ is a right inverse of $\beta$, then we define $\gamma : H \cdot G \to G'$ by $\gamma(h, g) = \alpha(h)\beta'(g)$. One easily checks that $\gamma$ is an isomorphism.

Conversely, given $\gamma : H \cdot G \to G'$, define $\beta' : G \to G'$ by $\beta'(g) = \gamma(1, g)$. Then one easily checks that $\beta'$ is a right inverse of $\beta$, so the sequence splits. This proves Lemma 1.2. □

**Exercise 1.3** Show that $\mathrm{Aut}(\Gamma_C)$ contains a subgroup isomorphic to the semi-direct product of $(\mathbb{Z}/2\mathbb{Z})^n$ (acting by sign changes) and $\mathrm{Aut}(C)$.

## 1.4 Root Lattices

We have seen that the lattice $\Gamma_{\widetilde{H}}$, for $\widetilde{H}$ the extended Hamming code, is generated by vectors $x \in \mathbb{R}^n$ with $x^2 = 2$. In this section we want to consider lattices with this property.

Let $\Gamma \subset \mathbb{R}^n$ be an even lattice, i.e., $x^2 \in 2\mathbb{Z}$ for all $x \in \Gamma$. Let

$$R := \{x \in \Gamma \mid x^2 = 2\}.$$

An element $x \in R$ is also called a *root*.

**Definition.** An even lattice $\Gamma \subset \mathbb{R}^n$ is called a *root lattice*, if $R$ generates $\Gamma$.

Now let $\Gamma$ be a root lattice, and let $R$ be the set of roots. Let $x, y \in R$. Then the Cauchy-Schwarz inequality

$$(x \cdot y)^2 \leq x^2 y^2 = 4$$

yields

$$x \cdot y \in \{0, +1, -1, +2, -2\}$$

and

$$x \cdot y = \pm 2 \iff x = \pm y.$$

Note that

$$x \cdot y = 1 \iff (x-y)^2 = 2 \iff x - y \in R.$$

**Theorem 1.1.** *Let $\Gamma \subset \mathbb{R}^n$ be a root lattice. Then $\Gamma$ contains a basis $(e_1, \ldots, e_n)$ with*

$$\begin{aligned} e_i^2 &= 2, \quad 1 \leq i \leq n, \\ e_i \cdot e_j &\in \{0, -1\}, \quad 1 \leq i, j \leq n, \ i \neq j. \end{aligned}$$

Theorem 1.1 goes back to Witt [94]. Note that $e_i \cdot e_j = -1$ means that the angle $\angle(e_i, e_j)$ between $e_i$ and $e_j$ is $120^o$.

**Definition.** A subset $S$ of $R$ is called a *fundamental system of roots* if

(i) $S$ is a basis of $\Gamma$.

(ii) Each $\beta \in R$ can be written as a linear combination $\beta = \sum_{\alpha \in S} k_\alpha \alpha$ with integral coefficients $k_\alpha$ all non-negative or all non-positive.

Let $S$ be a fundamental system of roots of $R$. Then $S$ is a basis as in Theorem 1.1: Let $\alpha, \beta \in S$, $\alpha \neq \beta$. If $\alpha \cdot \beta > 0$ then we must have $\alpha \cdot \beta = 1$. But this means that $\alpha - \beta$ is a root, contradicting condition (ii). Therefore $\alpha \cdot \beta \leq 0$ for all $\alpha, \beta \in S$ with $\alpha \neq \beta$.

In order to prove Theorem 1.1 it therefore suffices to show that each root system contains a fundamental system of roots. We follow [82] and [39] in the construction of such a fundamental system of roots.

Let $t \in \mathbb{R}^n$ be an element such that $t \cdot \alpha \neq 0$ for all $\alpha \in R$. Such an element exists: The set $R$ is finite, since it is the intersection of a discrete set with a compact set. The set

$$H_\alpha = \{x \in \mathbb{R}^n \mid x \cdot \alpha = 0\}$$

for $\alpha \in R$ is a hyperplane in $\mathbb{R}^n$. But the union of finitely many hyperplanes in $\mathbb{R}^n$ cannot exhaust $\mathbb{R}^n$, giving the existence of $t$. Let

$$R_t^+ = \{\alpha \in R \mid t \cdot \alpha > 0\}.$$

Then we have $R = R_t^+ \cup (-R_t^+)$. An element $\alpha \in R_t^+$ is called *decomposable*, if there exist $\beta, \gamma \in R_t^+$ such that $\alpha = \beta + \gamma$; otherwise, $\alpha$ is called *indecomposable*. Let $S_t$ be the set of indecomposable elements of $R_t^+$.

**Proposition 1.4.** *The set $S_t$ is a fundamental system of roots for $\Gamma$. Conversely, if $S$ is a fundamental system of roots, then there exists a $t \in \mathbb{R}^n$ such that $t \cdot \alpha > 0$ for all $\alpha \in S$, and $S = S_t$.*

For the proof of Proposition 1.4 we shall proceed in steps.

**Lemma 1.3.** *Each element of $R_t^+$ is a linear combination of elements of $S_t$ with non-negative integer coefficients.*

*Proof.* Otherwise, there would be an $\alpha \in R_t^+$ which does not have this property. Since $t \cdot \beta > 0$ for all $\beta \in R_t^+$, we may assume that $\alpha$ is chosen so that $t \cdot \alpha$ is minimal. The element $\alpha$ is decomposable, because otherwise it would belong to $S_t$. Therefore $\alpha = \beta + \gamma$ with $\beta, \gamma \in R_t^+$, and we have

$$t \cdot \alpha = t \cdot \beta + t \cdot \gamma.$$

Since $t \cdot \beta > 0$ and $t \cdot \gamma > 0$, $t \cdot \beta < t \cdot \alpha$ and $t \cdot \gamma < t \cdot \alpha$. Since $t \cdot \alpha$ was minimal, the elements $\beta$ and $\gamma$ can be written as linear combinations of elements of $S_t$ with non-negative integer coefficients, and hence $\alpha$, too. This contradicts the assumption. □

**Lemma 1.4.** $\alpha \cdot \beta \leq 0$ *for* $\alpha, \beta \in S_t$, $\alpha \neq \beta$.

*Proof.* Otherwise, $\alpha \cdot \beta = 1$ and hence $\gamma = \alpha - \beta \in R$. Then either $\gamma \in R_t^+$ or $-\gamma \in R_t^+$. In the first case $\alpha = \beta + \gamma$ would be decomposable, in the latter case $\beta = \alpha + (-\gamma)$ would be decomposable, both being contradictions. □

**Lemma 1.5.** *The elements of* $S_t$ *are linearly independent.*

*Proof.* Suppose that $\sum\limits_{\alpha \in S_t} a_\alpha \alpha = 0$. This relation can be written as

$$\sum b_\beta \beta = \sum c_\gamma \gamma$$

where $b_\beta, c_\gamma > 0$, and where $\beta$ and $\gamma$ range over disjoint subsets of $S_t$. Put $\lambda = \sum b_\beta \beta$. Then

$$\lambda \cdot \lambda = \sum_{\beta, \gamma} b_\beta c_\gamma (\beta \cdot \gamma) \leq 0$$

by Lemma 1.4, forcing $\lambda = 0$. Then

$$0 = t \cdot \lambda = \sum b_\beta (t \cdot \beta),$$

so that $b_\beta = 0$ for all $\beta$, and similarly $c_\gamma = 0$ for all $\gamma$, as required. □

**Lemma 1.6.** *Let* $\{\gamma_1, \ldots, \gamma_n\}$ *be a basis of* $\mathbb{R}^n$. *Then there exists an element* $t \in \mathbb{R}^n$ *with* $t \cdot \gamma_i > 0$ *for all* $1 \leq i \leq n$.

*Proof.* Let $\delta_i'$ be the orthogonal projection of $\gamma_i$ on the subspace $U_i$ spanned by all basis vectors except $\gamma_i$, for $i = 1, \ldots, n$. Let $\delta_i := \gamma_i - \delta_i'$. Then $\delta_i \cdot \gamma_j = 0$ for $1 \leq j \leq n$, $j \neq i$, and $\delta_i \cdot \gamma_i = \delta_i \cdot (\delta_i + \delta_i') = \delta_i^2 > 0$. Then $t = \sum\limits_{i=1}^{n} r_i \delta_i$ is an element as required provided that all $r_i > 0$. □

*Proof of Proposition 1.4.* That $S_t$ is a basis for $\Gamma$ as required follows from Lemmas 1.3, 1.5. Conversely, let $S$ be a fundamental system of roots of $R$, and let $t \in \mathbb{R}^n$ be such that $t \cdot \alpha > 0$ for all $\alpha \in S$. Such an element exists according to Lemma 1.6. Let $R^+$ denote the set of all roots which are linear combinations with non-negative integer coefficients of elements of $S$. Then we have $R^+ \subset R_t^+$ and $(-R^+) \subset (-R_t^+)$, so that $R^+ = R_t^+$ since $R = R^+ \cup (-R^+)$. Therefore the elements of $S$ are indecomposable in $R_t^+$, and hence $S \subset S_t$. Since $S$ and $S_t$ have the same number of elements, we have $S = S_t$. This finishes the proof of Proposition 1.4. □

It turns out that the possible systems $\{e_1,\ldots,e_n\}$ with the properties of Theorem 1.1 can be *classified*. We recall that we can associate a graph (the *Coxeter-Dynkin diagram*) to $\{e_1,\ldots,e_n\}$. We represent each $e_i$ by a vertex. The edges are defined by the following rule ($i \neq j$)

$$e_i \cdot e_j = -1 \Longleftrightarrow \overset{\bullet}{e_i} \text{———} \overset{\bullet}{e_j}$$

$$e_i \cdot e_j = 0 \Longleftrightarrow \overset{\bullet}{e_i} \qquad \overset{\bullet}{e_j}$$

Let $G$ be the Coxeter-Dynkin diagram corresponding to a basis $\{e_1,\ldots,e_n\}$ as in Theorem 1.1.

**Lemma 1.7.** *$G$ contains no cycles.*

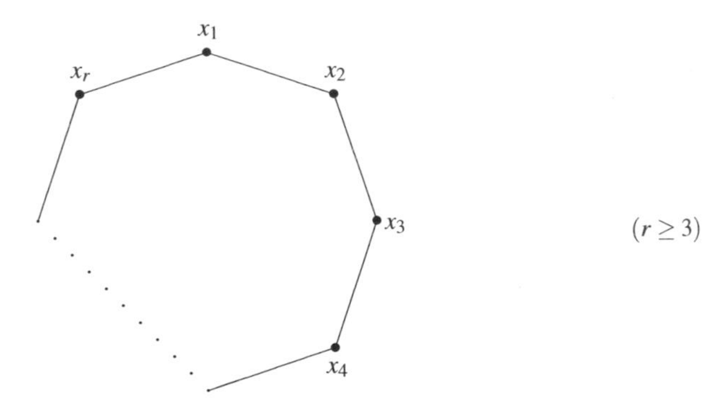


**Fig. 1.4** A minimal cycle

*Proof.* Suppose that $G$ contains a minimal cycle as in Fig. 1.4, where $x_1,\ldots,x_r$ denote the corresponding basis vectors. Then we have

$$(x_1+\ldots+x_r)^2 = 2r-2r = 0,$$

which is impossible. This proves Lemma 1.7. □

**Lemma 1.8.** *$G$ does not contain a subgraph of the form of Fig. 1.5.*

*Proof.* For the basis vectors corresponding to the subgraph of Fig. 1.5, we have

$$(2x_1+\ldots+2x_r+x_{r+1}+\ldots+x_{r+4})^2 = 8r+8-8(r-1)-16 = 0.$$

This proves Lemma 1.8. □

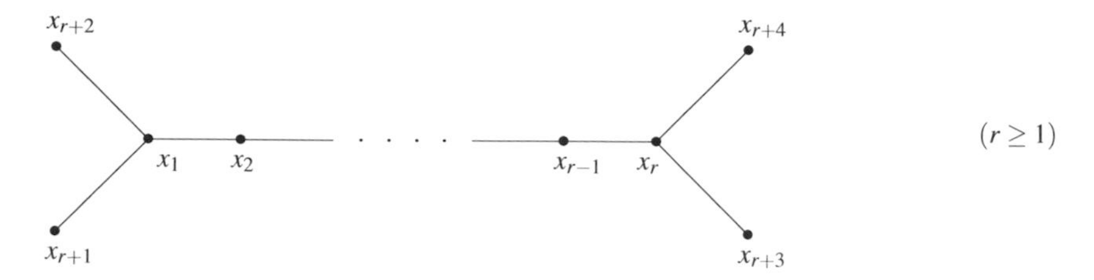


**Fig. 1.5** A graph not contained in a Coxeter-Dynkin diagram of a root lattice

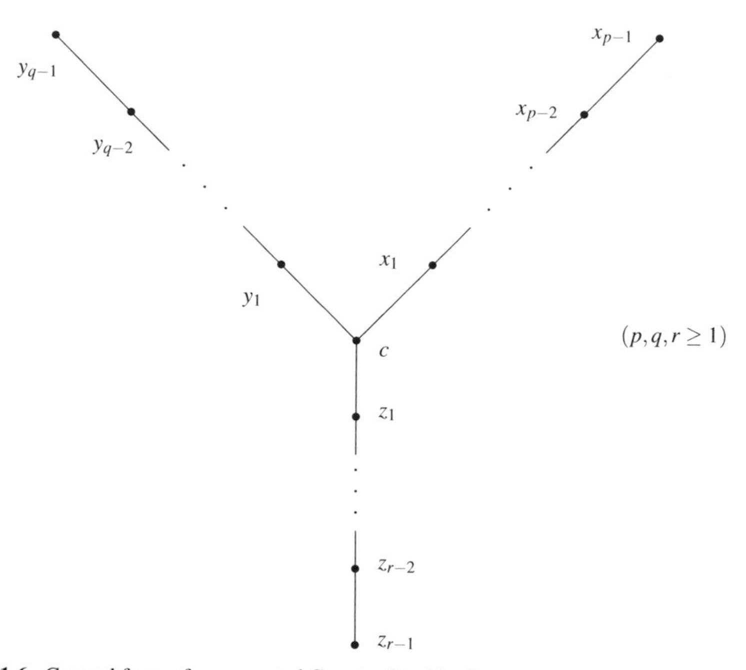


**Fig. 1.6** General form of a connected Coxeter-Dynkin diagram

Now suppose that $G$ is connected. Then according to Lemmas 1.7 and 1.8 the graph must have the shape of Fig. 1.6. Let

$$\begin{aligned} w = c &+ \frac{1}{p}[(p-1)x_1 + (p-2)x_2 + \cdots + x_{p-1}] \\ &+ \frac{1}{q}[(q-1)y_1 + (q-2)y_2 + \cdots + y_{q-1}] \\ &+ \frac{1}{r}[(r-1)z_1 + (r-2)z_2 + \cdots + z_{r-1}]. \end{aligned}$$

Then the vectors $x_1,\dots,x_{p-1},y_1,\dots,y_{q-1},z_1,\dots,z_{r-1},w$ form a basis of the $\mathbb{Q}$-vector space $\Gamma\otimes\mathbb{Q}$ over $\mathbb{Q}$. Note that $x_i\perp y_j$, $x_i\perp z_j$, $y_i\perp z_j$, for all $i$, $j$, and $w\perp x_i$, $w\perp y_i$, $w\perp z_i$, for all $i$, since, for example,

$$\begin{aligned} w\cdot x_1 &= c\cdot x_1 + \frac{(p-1)\cdot 2-(p-2)}{p} = 0, \\ w\cdot x_i &= \frac{p-i}{p}\cdot 2 - \frac{p-i+1}{p} - \frac{p-i-1}{p} = 0,\, p-2\geq i\geq 2, \\ w\cdot x_{p-1} &= \frac{1}{p}(2-(p-(p-2))) = 0. \end{aligned}$$

Therefore

$$\begin{aligned} w^2 = w\cdot c &= 2-\frac{p-1}{p}-\frac{q-1}{q}-\frac{r-1}{r} \\ &= -1+\frac{1}{p}+\frac{1}{q}+\frac{1}{r}. \end{aligned}$$

Since the scalar product is positive definite, $w^2>0$. This yields the following necessary condition on the numbers $p,q,r$:

$$\frac{1}{p}+\frac{1}{q}+\frac{1}{r}>1.$$

This inequality has the following solutions with $p\leq q\leq r$:

$$\begin{aligned} (p,q,r) &= (1,q,r),\ q\leq r \text{ arbitrary}, \\ & \quad (2,2,r),\ 2\leq r \text{ arbitrary}, \\ & \quad (2,3,3), \\ & \quad (2,3,4), \\ & \quad (2,3,5). \end{aligned}$$

A lattice $\Gamma$ is called *reducible*, if $\Gamma$ is the orthogonal direct sum $\Gamma=\Gamma_1\perp\Gamma_2$ of two lattices $\Gamma_1\subset\mathbb{R}^{n_1}$, $\Gamma_2\subset\mathbb{R}^{n_2}$ with $n_1,n_2\geq 1$; otherwise it is called *irreducible*. If the Coxeter-Dynkin diagram $G$ corresponding to a root lattice $\Gamma$ is connected, then the lattice $\Gamma$ is irreducible. In general the root lattice $\Gamma$ is an orthogonal direct sum of lattices $\Gamma_i$ corresponding to the connected components $G_i$ of the graph $G$. Therefore we have proved the following theorem up to the existence of the corresponding lattices.

**Theorem 1.2.** *Every root lattice is the orthogonal direct sum of the irreducible root lattices with the Coxeter-Dynkin diagrams indicated in Fig. 1.7.*

We now show the *existence* of the corresponding irreducible lattices.

$A_n$: Consider $\mathbb{R}^{n+1}$ with standard basis $(\varepsilon_1,\dots,\varepsilon_{n+1})$. Let $e$ be the vector $e=\varepsilon_1+\varepsilon_2+\dots+\varepsilon_{n+1}=(1,1,\dots,1)\in\mathbb{R}^{n+1}$. Let

$$\Gamma=\{x=(x_1,\dots,x_{n+1})\in\mathbb{Z}^{n+1}\mid x\cdot e=x_1+\dots+x_{n+1}=0\}.$$

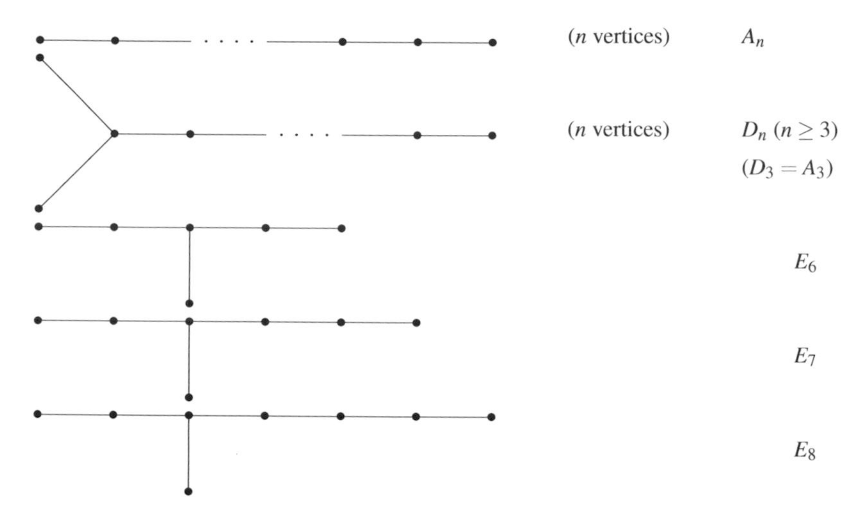


**Fig. 1.7** Coxeter-Dynkin diagrams of the irreducible root lattices

Then $\Gamma$ is a lattice in $e^{\perp} \cong \mathbb{R}^n$. The lattice $\Gamma$ is generated by elements $\varepsilon_i - \varepsilon_j$ $(1 \leq i, j \leq n+1,\, i \neq j)$, which are the roots in $\Gamma$. There are $n(n+1)$ of them. The set $\{\varepsilon_2 - \varepsilon_1, \varepsilon_3 - \varepsilon_2, \ldots, \varepsilon_{n+1} - \varepsilon_n\}$ is a basis of $\Gamma$ with Coxeter-Dynkin diagram

$\varepsilon_2 - \varepsilon_1$ $\quad$ $\varepsilon_3 - \varepsilon_2$ $\quad$ $\varepsilon_4 - \varepsilon_3$ $\quad \cdots \quad$ $\varepsilon_{n+1} - \varepsilon_n$

Let $A$ be the matrix of the scalar product with respect to this basis (cf. Sect. 1.1). Then

$$A = \begin{pmatrix} 2 & -1 & & & 0 \\ -1 & 2 & -1 & & \\ & -1 & 2 & \ddots & \\ & & \ddots & \ddots & -1 \\ 0 & & & -1 & 2 \end{pmatrix}.$$

Then one easily computes

$$\operatorname{disc}(A_n) = \det A = n+1\,.$$

$D_n$: For $n \geq 3$ let

$$\Gamma = \{(x_1, \ldots, x_n) \in \mathbb{Z}^n \mid x_1 + \ldots + x_n \text{ even}\}$$

In other words, $\Gamma$ is obtained by coloring the points of the cubic lattice $\mathbb{Z}^n$ alternately black and white with a checkerboard coloring, and taking the black points. Then $\Gamma$ is a lattice in $\mathbb{R}^n$, and it is generated by elements $\pm\varepsilon_i \pm \varepsilon_j$ and $\pm\varepsilon_i \mp \varepsilon_j$ $(1 \leq i, j \leq n,\, i \neq j)$, which are the roots in $\Gamma$. There are $2n(n-1)$ of them. The set $\{\varepsilon_2 - \varepsilon_1, \varepsilon_3 - \varepsilon_2, \ldots, \varepsilon_n - \varepsilon_{n-1}, \varepsilon_1 + \varepsilon_2\}$ is a basis of $\Gamma$ with Coxeter-Dynkin diagram

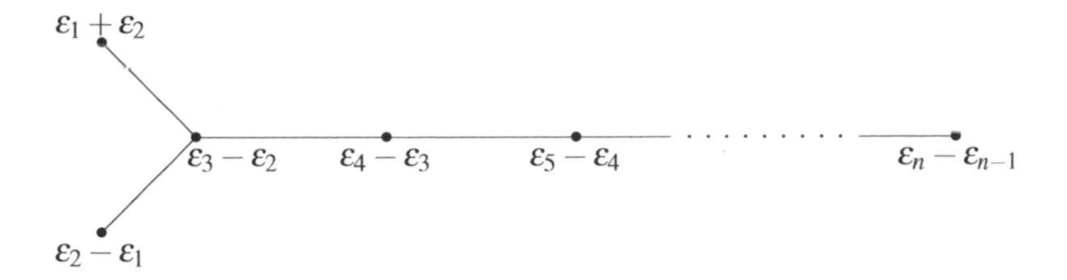


The lattice $\Gamma$ is of index 2 in the lattice $\mathbb{Z}^n$, so

$$\text{disc}\,(D_n) = 4.$$

To give another construction of $D_n$, consider the even-weight-code

$$C = \{(u_1, \ldots, u_n) \in \mathbb{F}_2^n \mid u_1 + \ldots + u_n = 0\}.$$

Then $\Gamma = \rho^{-1}(C)$, where

$$\rho : \mathbb{Z}^n \longrightarrow (\mathbb{Z}/2\mathbb{Z})^n$$

is the reduction-mod-2-mapping. The code $C$ is the kernel of the linear function

$$\begin{array}{rl} \mathbb{F}_2^n & \longrightarrow \mathbb{F}_2 \\ (u_1, \ldots, u_n) & \longmapsto u_1 + \ldots + u_n \end{array}$$

Therefore

$$\mathbb{Z}^n/\rho^{-1}(C) \cong \mathbb{F}_2^n/C \cong \mathbb{Z}/2\mathbb{Z}.$$

This also yields $\text{disc}\,(D_n) = 4$.

$E_8$: We have already constructed this lattice from the extended Hamming code of length 8. We know that this lattice is unimodular.

$E_7$: Consider the lattice $E_8$, and let $(e_1, \ldots, e_8)$ be a fundamental system of roots with Coxeter-Dynkin diagram (cf. Sect. 1.3)

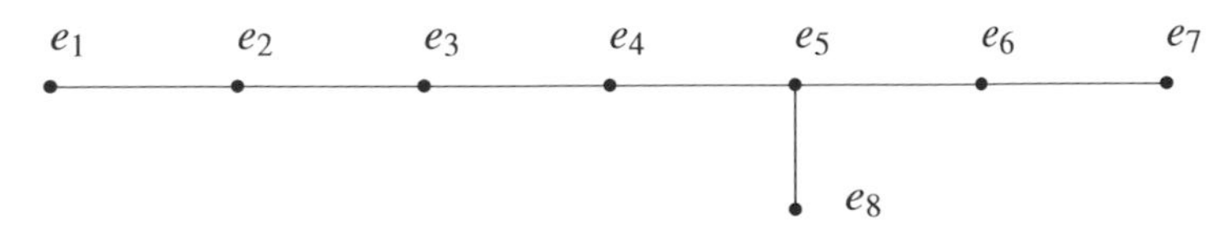


Let $v := 2e_1 + 3e_2 + 4e_3 + 5e_4 + 6e_5 + 4e_6 + 2e_7 + 3e_8$. Then $v$ is a root of $E_8$. Define

$$\Gamma = \{x \in E_8 \mid x \cdot v = 0\}.$$

This is a lattice in the orthogonal complement of the vector $v$ in $\mathbb{R}^8$ which is isomorphic to $\mathbb{R}^7$. Since $e_i \cdot v = 0$ for $i = 2, \ldots, 8$, the fundamental roots $e_2, \ldots, e_8$ of $E_8$ are contained in $\Gamma$. The Coxeter-Dynkin diagram corresponding to these roots is a graph of type $E_7$. Using $e_1 \cdot v = 1$, one can easily see that these roots generate $\Gamma$. Hence $\Gamma$ is a root lattice of type $E_7$. By Proposition 1.2 we have

$$\text{disc}(E_7) = \text{disc}(A_1) = 2.$$

$E_6$: Consider again the lattice $E_8$ and let $v_1 = v$ and $v_2 = -e_1$. Then $v_1 \cdot v_2 = -1$. Define

$$\Gamma = \{x \in E_8 \mid x \cdot v_1 = x \cdot v_2 = 0\}.$$

This is a lattice in the orthogonal complement of the two vectors $v_1$ and $v_2$ in $\mathbb{R}^8$ which is isomorphic to $\mathbb{R}^6$. Since $e_i \cdot v_2 = 0$ for $i = 3, \ldots, 8$, the fundamental roots $e_3, \ldots, e_8$ of $E_8$ are contained in $\Gamma$. The Coxeter-Dynkin diagram corresponding to these roots is a graph of type $E_6$. Again one can easily see that these roots generate $\Gamma$. Hence $\Gamma$ is a root lattice of type $E_6$. By Proposition 1.2 we have

$$\text{disc}(E_6) = \text{disc}(A_2) = 3.$$

Therefore we have shown the existence of lattices corresponding to the Coxeter-Dynkin diagrams of type $A_n$, $D_n$, $E_6$, $E_7$ and $E_8$. This completes the proof of Theorem 1.2.

We have constructed the lattice $E_8$ as the lattice $\Gamma_C$ for a suitable binary linear code $C$. We now want to discuss the question: For which of the irreducible root lattices $\Gamma$ do there exist binary linear codes $C$ with $\Gamma = \Gamma_C$? Note that until now we only know that $\Gamma = \Gamma_C$ for $E_8$, and in case $D_n$ we know $\Gamma = \rho^{-1}(C)$, but not $\Gamma = \Gamma_C = \frac{1}{\sqrt{2}}\rho^{-1}(C)$. The following proposition will give a precise answer to this question. In order to state this proposition, we need another definition.

Let $\Gamma$ be a root lattice in $\mathbb{R}^n$, and $\alpha \in \Gamma$ be a root. Consider the linear transformation of $\mathbb{R}^n$ given by

$$x \longmapsto x - 2\frac{x \cdot \alpha}{\alpha \cdot \alpha}\alpha = x - (x \cdot \alpha)\,\alpha$$

This is a *reflection* at the hyperplane $\alpha^\perp$ orthogonal to $\alpha$. We call it $s_\alpha$. It leaves $\alpha^\perp$ pointwise fixed, and sends any vector orthogonal to the hyperplane $\alpha^\perp$ into its negative. Let $W(\Gamma)$ be the subgroup of $\mathrm{GL}_n(\mathbb{R})$ generated by the reflections $s_\alpha$ corresponding to the roots $\alpha$ of $\Gamma$. This group is called the *Weyl group* of $\Gamma$.

For the following two lemmas see also [39, 10.4].

**Lemma 1.9.** *Let $\Gamma \subset \mathbb{R}^n$ be an irreducible root lattice. Then the Weyl group $W(\Gamma)$ acts irreducibly on $\mathbb{R}^n$, i.e., if $U$ is a $W(\Gamma)$-invariant subspace of $\mathbb{R}^n$, then either $U = 0$, or $U = \mathbb{R}^n$.*

*Proof.* Let $U$ be a nonzero subspace of $\mathbb{R}^n$ invariant under $W(\Gamma)$. The orthogonal complement $U^\perp$ of $U$ in $\mathbb{R}^n$ is also $W(\Gamma)$-invariant, and $\mathbb{R}^n = U \oplus U^\perp$. Let $\alpha \in \Gamma$ be a root. Suppose that $\alpha \notin U$. Let $u \in U$ be arbitrary. Since $s_\alpha(U) = U$, $u - (u \cdot \alpha)\alpha \in U$. This implies that $u \cdot \alpha = 0$. Since this is true for any $u \in U$, we have $\alpha \in U^\perp$. Therefore each root lies in $U$ or in $U^\perp$. Since $R$ spans $\Gamma$, we have $\Gamma = (U \cap \Gamma) \perp (U^\perp \cap \Gamma)$. But this implies that $U^\perp \cap \Gamma = \{0\}$, since $\Gamma$ is irreducible. Therefore $\Gamma = U \cap \Gamma$, and hence $U = \mathbb{R}^n$, since $\Gamma$ spans $\mathbb{R}^n$. This proves Lemma 1.9. □

One says that a group $G$ acts *transitively* on a set $X$ if for any $x, y \in X$ there exists a $g \in G$ such that $\phi(g, x) = y$ where $\phi : G \times X \to X$ denotes the group action.

**Lemma 1.10.** *Let $\Gamma \subset \mathbb{R}^n$ be an irreducible root lattice. Then the Weyl group $W(\Gamma)$ acts transitively on the set of roots $R$.*

*Proof.* Let $\alpha$ and $\beta$ be arbitrary roots of $\Gamma$. By Lemma 1.9, the elements $w(\alpha)$, $w \in W(\Gamma)$, span $\mathbb{R}^n$. Therefore not all $w(\alpha)$ can be orthogonal to $\beta$. After replacing $\alpha$ by $w(\alpha)$ for some $w \in W(\Gamma)$ if necessary, we may assume that $\alpha$ and $\beta$ are non-orthogonal. We may also assume that $\alpha$ and $\beta$ are distinct and $\alpha \neq -\beta$, because otherwise we are done. Hence $\alpha \cdot \beta = \pm 1$. Replacing $\beta$ if necessary by $-\beta = s_\beta(\beta)$, we may assume that $\alpha \cdot \beta = 1$. Then $s_\alpha s_\beta s_\alpha(\beta) = s_\alpha s_\beta(\beta - \alpha) = s_\alpha(\beta - \alpha - \beta) = \alpha$. This proves Lemma 1.10. □

**Remark 1.1** Lemma 1.10 implies that in the constructions for the lattices $E_7$ and $E_6$ above one can take any root $v$ and any pair of roots $v_1$, $v_2$ with $v_1 \cdot v_2 = -1$ respectively.

**Proposition 1.5.** *Let $\Gamma \subset \mathbb{R}^n$ be an irreducible root lattice. Then the following statements are equivalent:*
(i) *$\Gamma = \Gamma_C$ for a binary linear code $C \subset \mathbb{F}_2^n$.*
(ii) *$\Gamma$ contains $n$ pairwise orthogonal roots.*
(iii) *$nA_1 = A_1 \perp \ldots \perp A_1$ ($n$ times) $\subset \Gamma$.*
(iv) *$-1 \in W(\Gamma)$.*
(v) *$2\Gamma^* \subset \Gamma$.*
(vi) *$\Gamma$ is of type $A_1$, $D_n$ ($n \geq 4$, $n$ even), $E_7$, or $E_8$.*

*Proof.* (i)⇒(ii): Let $\Gamma = \Gamma_C$ for a binary code $C \subset \mathbb{F}_2^n$, and let $(\varepsilon_1, \ldots, \varepsilon_n)$ be the standard basis of $\mathbb{R}^n$. Then $\frac{1}{\sqrt{2}} 2\varepsilon_i \in \Gamma_C$ for all $1 \leq i \leq n$. These are pairwise orthogonal roots.

(ii)⇔(iii): This is trivial, because a single root spans a lattice of type $A_1$.

(ii)⇒(iv): Let $\alpha_1, \ldots, \alpha_n$ be pairwise orthogonal roots of $\Gamma$. Then $\alpha_1, \ldots, \alpha_n$ form an $\mathbb{R}$-basis of the vector space $\mathbb{R}^n$. Let $x \in \mathbb{R}^n$. Then $x$ can be written $x = \sum_i \xi_i \alpha_i$ with $\xi_i \in \mathbb{R}$. Then

$$\begin{aligned} s_{\alpha_1} s_{\alpha_2} \ldots s_{\alpha_n}(x) &= s_{\alpha_1} s_{\alpha_2} \ldots s_{\alpha_n}\left(\sum_i \xi_i \alpha_i\right) \\ &= -\sum_i \xi_i \alpha_i \\ &= -x, \end{aligned}$$

so $-1 = s_{\alpha_1} s_{\alpha_2} \ldots s_{\alpha_n} \in W(\Gamma)$.

(iv)⇒(v): Let $-1 \in W(\Gamma)$. Then $-1 = s_{\beta_1} s_{\beta_2} \ldots s_{\beta_k}$ for certain roots $\beta_1, \ldots, \beta_k \in \Gamma$. Let $x \in \Gamma^*$. Then $-x = s_{\beta_1} s_{\beta_2} \ldots s_{\beta_k}(x) = x + y$ with $y \in \Gamma$. Therefore

$$2x = -y \in \Gamma.$$

(v)⇔(vi): The quotient $\Gamma^*/\Gamma$ is a finite abelian group of order $|\Gamma^*/\Gamma| = \operatorname{disc}(\Gamma)$. Now $2\Gamma^* \subset \Gamma$ is equivalent to

$$\Gamma^*/\Gamma \cong (\mathbb{Z}/2\mathbb{Z})^l, \ \operatorname{disc}(\Gamma) = 2^l,$$

for some $l \geq 0$. For $\Gamma = A_n$

$$\Gamma = \left\{ (x_1, x_2, \ldots, x_{n+1}) \in \mathbb{Z}^{n+1} \;\middle|\; \sum_i x_i = 0 \right\},$$

$\Gamma^*$ is generated by $\Gamma$ and $\left(\frac{n}{n+1}, -\frac{1}{n+1}, -\frac{1}{n+1}, \ldots, -\frac{1}{n+1}\right)$, and hence

$$\Gamma^*/\Gamma \cong \mathbb{Z}/(n+1)\mathbb{Z}.$$

For $\Gamma = D_n$

$$\Gamma = \left\{ (x_1, \ldots, x_n) \in \mathbb{Z}^n \;\middle|\; \sum_i x_i \text{ even} \right\},$$

$$\Gamma^* = \mathbb{Z}^n + \mathbb{Z}\left(\frac{1}{2}, \frac{1}{2}, \ldots, \frac{1}{2}\right),$$

and

$$\Gamma^*/\Gamma \cong \begin{cases} (\mathbb{Z}/2\mathbb{Z}) \times (\mathbb{Z}/2\mathbb{Z}) & \text{for } n \text{ even}, \\ (\mathbb{Z}/4\mathbb{Z}) & \text{for } n \text{ odd}, \end{cases}$$

where $\Gamma^*/\Gamma$ is generated by $\omega_1 = (1, 0, \ldots, 0)$ and $\omega_2 = (\frac{1}{2}, \frac{1}{2}, \ldots, \frac{1}{2})$, if $n$ is even, and by $\omega_2$, if $n$ is odd. Note that $\omega_1 = 2\omega_2 \pmod{\Gamma}$ if $n$ is odd. Thus $2\Gamma^* \subset \Gamma$ if and only if $\Gamma = A_1$, $D_n$ ($n \geq 4$, $n$ even), $E_7$, or $E_8$.

(vi)⇒(i): We indicate a binary linear code $C$ with $\Gamma_C = \Gamma$ in each of the listed cases:

$A_1$: $C = \{0\} \subset \mathbb{F}_2$

$D_n$ ($n \geq 4$, $n$ even): Here $C$ is the "double" of the even weight code $\widetilde{C} \subset \mathbb{F}_2^{n/2}$, i.e.,

$$\widetilde{C} = \left\{ (u_1, \ldots, u_{n/2}) \in \mathbb{F}_2^{n/2} \mid \sum u_i = 0 \right\} \subset \mathbb{F}_2^{n/2},$$

$$C = \{ v \in \mathbb{F}_2^n \mid v = (u_1, u_1, \ldots, u_{n/2}, u_{n/2}),\, u = (u_1, u_2, \ldots, u_{n/2}) \in \widetilde{C} \}.$$

$E_7$: $C = H^\perp \subset \mathbb{F}_2^7$, where $H$ is the Hamming code. This code consists of 0 and the 7 codewords of weight 4 of the Hamming code $H$.

$E_8$: $C = \widetilde{H} \subset \mathbb{F}_2^8$, where $\widetilde{H}$ is the extended Hamming code, as we have seen above.

This completes the proof of Proposition 1.5. □

**Exercise 1.4** Let $\Gamma \subset \mathbb{R}^n$ be a root lattice and $w \in W(\Gamma)$ be an involution. Show that there exists a set $S$ of pairwise orthogonal roots such that $w$ is equal to the product of the $s_\alpha$, $\alpha \in S$.

We give a second, direct, proof of (i)⇒(v) in Proposition 1.5. For this proof we need a general fact which we formulate as a lemma.

**Lemma 1.11.** *Let $C \subset \mathbb{F}_2^n$ be a binary linear code. Then*

$$\Gamma_C^* = \Gamma_{C^\perp}.$$

*Proof.* Let $\rho : \mathbb{Z}^n \longrightarrow \mathbb{F}_2^n$ be the mapping induced by reduction mod 2. Let $y \in \Gamma_{C^\perp}$, $x \in \Gamma_C$. Then $y = \frac{1}{\sqrt{2}}\widetilde{y}$, $x = \frac{1}{\sqrt{2}}\widetilde{x}$ with $\widetilde{y} \in \rho^{-1}(C^\perp)$, $\widetilde{x} \in \rho^{-1}(C)$. Now $\widetilde{x} \cdot \widetilde{y} \in 2\mathbb{Z}$, and therefore

$x \cdot y \in \mathbb{Z}$. This proves $\Gamma_{C^\perp} \subset \Gamma_C^*$.

Now let $k = \dim C$. Then

$$\mu = \operatorname{vol}(\mathbb{R}^n/\Gamma_C) = \frac{2^{n-k}}{2^{n/2}} = 2^{\frac{n}{2}-k},$$
$$\mu^* = \operatorname{vol}(\mathbb{R}^n/\Gamma_C^*) = \frac{1}{\mu} = 2^{k-\frac{n}{2}} = \operatorname{vol}(\mathbb{R}^n/\Gamma_{C^\perp}),$$

which implies that $\Gamma_{C^\perp} = \Gamma_C^*$. □

*Proof of Proposition 1.5*, (i)⇒(v): Let $\Gamma = \Gamma_C$ for a binary linear code $C \subset \mathbb{F}_2^n$. Then $\Gamma^* = \Gamma_{C^\perp}$ by Lemma 1.11. Let $x \in \Gamma_C^*$. Then $x = \frac{1}{\sqrt{2}}(c + 2y)$ for $c \in C^\perp$ and $y \in \mathbb{Z}^n$. Then $2x = \frac{1}{\sqrt{2}}(0 + 2z)$, with $z \in \mathbb{Z}^n$. Thus $2x \in \Gamma_C$, what had to be shown. □

In Table 1.1 we have listed the groups $\Gamma^*/\Gamma$ and their orders in each case. We have also listed the numbers of roots $|R|$. These are obtained as follows: for $A_n$ and $D_n$ see the existence proof concerning Theorem 1.2. The number of roots of $E_8$ was computed in Sect. 1.3. Similarly one gets the number of roots of $E_7$, using that $E_7 = \Gamma_{H^\perp}$ where $H$ is the Hamming code of length 7. For $E_6$ see Example 5.1 and Exercise 5.1.

**Exercise 1.5** Determine the number of roots of $E_7$, using $E_7 = \Gamma_{H^\perp}$.

If $\Gamma \subset \mathbb{R}^n$ is a root lattice and $R$ is its set of roots then the number

$$h := \frac{|R|}{n}$$

is called the *Coxeter number* of $\Gamma$. The Coxeter numbers of the irreducible root lattices are listed in the last column of Table 1.1.

**Table 1.1** List of groups $\Gamma^*/\Gamma$

| $\Gamma$ | $\Gamma^*/\Gamma$ | $\lvert\Gamma^*/\Gamma\rvert$ | $\lvert R\rvert$ | $h$ |
|---|---|---|---|---|
| $A_n$ | $\mathbb{Z}/(n+1)\mathbb{Z}$ | $n+1$ | $n(n+1)$ | $n+1$ |
| $D_n$ | $\begin{cases} (\mathbb{Z}/2\mathbb{Z}) \times (\mathbb{Z}/2\mathbb{Z}) & n \text{ even} \\ (\mathbb{Z}/4\mathbb{Z}) & n \text{ odd} \end{cases}$ | 4 | $2n(n-1)$ | $2(n-1)$ |
| $E_6$ | $\mathbb{Z}/3\mathbb{Z}$ | 3 | 72 | 12 |
| $E_7$ | $\mathbb{Z}/2\mathbb{Z}$ | 2 | 126 | 18 |
| $E_8$ | $\{0\}$ | 1 | 240 | 30 |

**Proposition 1.6.** *Let $\Gamma \subset \mathbb{R}^n$ be an irreducible root lattice. Then for a fixed $y \in \mathbb{R}^n$ we have*

$$\sum_{x \in R} (x \cdot y)^2 = 2 \cdot h \cdot y^2,$$

*where $h$ is the Coxeter number of $\Gamma$.*

For the proof of this proposition we need some preparations.

Let $P \in \mathbb{C}[y_1, \ldots, y_n]$ be a complex polynomial in $n$ variables $y_1, \ldots, y_n$. Such a polynomial $P$ is called *harmonic* or *spherical*, if and only if $\Delta P = 0$, where

$$\Delta = \sum_{i=1}^{n} \frac{\partial}{\partial y_i^2}$$

is the Laplace operator.

Let $\Gamma \subset \mathbb{R}^n$ be a root lattice and let $R$ be its set of roots. Consider the polynomial

$$f(y) := \sum_{x \in R} \left( (x \cdot y)^2 - \frac{1}{n} x^2 y^2 \right)$$

in the variables $y_1, \ldots, y_n$. Since

$$\begin{aligned} \Delta f &= \sum_{x \in R} \left( \Delta (x_1 y_1 + \ldots + x_n y_n)^2 - \frac{1}{n} \Delta \left( x^2 \left( y_1^2 + \ldots + y_n^2 \right) \right) \right) \\ &= \sum_{x \in R} \left( 2 \left( x_1^2 + \ldots + x_n^2 \right) - \frac{2n}{n} x^2 \right) \\ &= 0, \end{aligned}$$

the polynomial $f$ is harmonic.

We also need the following lemma (cf. [6, Chap. V, §2, 1, Proposition 1]).

**Lemma 1.12.** *Let $\Gamma \subset \mathbb{R}^n$ be an irreducible root lattice, and let $W(\Gamma)$ be the Weyl group of $\Gamma$. Then the following is true:*

(i) *Each endomorphism of $\mathbb{R}^n$ commuting with every element of $W(\Gamma)$ is a homothety (i.e., a scalar multiple of the identity).*

(ii) *Let $b$ be a nonzero bilinear form on $\mathbb{R}^n$ invariant under $W(\Gamma)$. Then there exists a $\rho \in \mathbb{R}$, $\rho \neq 0$, such that*

$$b(x, y) = \rho (x \cdot y)$$

*for all $x, y \in \mathbb{R}^n$.*

*Proof.* (i) Let $g$ be an endomorphism of $\mathbb{R}^n$ commuting with every element of $W(\Gamma)$. Let $\alpha \in \Gamma$ be a root, and let $L = \mathbb{R}\alpha$. Then $g(L) \subset L$, since

$$-g(x) = g(s_\alpha x) = s_\alpha g(x)$$

for all $x \in L$ and $L = \{x \in \mathbb{R}^n \mid s_\alpha x = -x\}$. Therefore there exists a $\rho \in \mathbb{R}$ such that $g(x) = \rho x$ for all $x \in L$. Let $U$ be the kernel of $g - \rho \cdot 1$ . Then $U$ is a subspace of $\mathbb{R}^n$ which is invariant under $W(\Gamma)$ and nonzero, since it contains $L$. By Lemma 1.9, $U = \mathbb{R}^n$, and $g = \rho \cdot 1$ . This proves (i).

(ii) Let $b$ be a nonzero bilinear form on $\mathbb{R}^n$ invariant under $W(\Gamma)$. From linear algebra, it is known that there exists an endomorphism $g$ of $\mathbb{R}^n$ such that

$$b(x, y) = (g(x)) \cdot y$$

for all $x,y \in \mathbb{R}^n$. Since $b$ is invariant under $W(\Gamma)$, the endomorphism $g$ commutes with all elements of $W(\Gamma)$. In fact, let $x,y \in \mathbb{R}^n$ and $w \in W(\Gamma)$. Then one has

$$\begin{aligned}(gw(x)) \cdot y &= b(w(x),y)\\ &= b\left(x,w^{-1}(y)\right)\\ &= (g(x)) \cdot \left(w^{-1}(y)\right)\\ &= (wg(x)) \cdot y,\end{aligned}$$

and hence $gw(x) = wg(x)$. By (i) there exists a $\rho \in \mathbb{R}$ such that $g = \rho \cdot 1$ . Hence

$$b(x,y) = \rho(x \cdot y)$$

for all $x,y \in \mathbb{R}^n$, which proves (ii). □

*Proof of Proposition 1.6.* The polynomial $f(y)$ is a quadratic form in $y_1,\ldots,y_n$ invariant under the Weyl group $W(\Gamma)$. By Lemma 1.12(ii) $f(y)$ is a multiple of the quadratic form $y^2$. Since $f$ is harmonic and $\Delta(y^2) = 2n$, $f$ has to be equal to zero. This yields the formula of Proposition 1.6. □

## 1.5 Highest Root and Weyl Vector

We continue our discussion of root lattices. The material in this section, however, will only be needed in Chapter 4. A general reference for this section is [6].

Let $\Gamma \subset \mathbb{R}^n$ be an irreducible root lattice, let $R$ be its set of roots, and let $(e_1,\ldots,e_n)$ be a fundamental system of roots of $\Gamma$. Then each root $\alpha \in R$ can be written as $\alpha = \sum_{i=1}^{n} k_i e_i$ with integral coefficients $k_i$ all non-negative or all non-positive. If all $k_i \geq 0$ (resp. all $k_i \leq 0$) we call $\alpha$ *positive* (resp. *negative*) and write $\alpha \geq 0$ (resp. $\alpha \leq 0$). The collections of positive and negative roots (relative to $(e_1,\ldots,e_n)$) will usually be denoted by $R^+$ and $R^-$ respectively. We define a partial ordering on the set $R$ of roots of $\Gamma$, compatible with the notation $\alpha \geq 0$, as follows. For $\alpha,\beta \in R$ we define $\alpha \geq \beta$ if and only if $\alpha - \beta = \sum k_i e_i$ with all $k_i \geq 0$.

**Lemma 1.13.** *Relative to the partial ordering $\leq$ on R, there is a unique maximal root $\beta$. This root satisfies the following two properties:*
(i) $\beta \cdot e_i \geq 0$ *for all* $1 \leq i \leq n$.
(ii) *If* $\beta = \sum m_i e_i$ *then all* $m_i > 0$.

*Proof.* Let $\beta = \sum m_i e_i$ be maximal in the ordering. Then $\beta \geq 0$ or $\beta \leq 0$. If $\beta \leq 0$, $\beta \neq 0$, then $\beta \leq -\beta$, $\beta \neq -\beta$, which is a contradiction to the maximality of $\beta$. Thus $m_i \geq 0$ for all $1 \leq i \leq n$. Let $I = \{i \mid 1 \leq i \leq n,\ m_i > 0\}$ and $J$ be the complement of $I$ in the set $\{1,2,\ldots,n\}$. Then $I \neq \emptyset$. Suppose $J$ is non-void. Since $(e_1,\ldots,e_n)$ is a fundamental system of roots, $e_i \cdot e_j \leq 0$ for all $1 \leq i,j \leq n$, and since $\Gamma$ is irreducible, there exists an $i \in I$ and a $j \in J$ with $e_i \cdot e_j < 0$. Then

$$\beta \cdot e_j = \sum_{k\in I} m_k (e_k \cdot e_j) < 0.$$

But this implies that $\beta + e_j$ is a root, contradicting the maximality of $\beta$. Therefore $J$ is empty and all $m_i > 0$. This argument also shows that $\beta \cdot e_i \geq 0$ for all $1 \leq i \leq n$, with $\beta \cdot e_i > 0$ for at least one $i$, since $\{e_1, \ldots, e_n\}$ spans $\Gamma$.

Now let $\beta'$ be another maximal root. The preceding argument applies to $\beta'$, so $\beta'$ involves (with positive coefficient) at least one $e_i$ with $\beta \cdot e_i > 0$. It follows that $\beta' \cdot \beta > 0$, and $\beta - \beta'$ is a root unless $\beta = \beta'$. But if $\beta - \beta'$ is a root, then either $\beta \leq \beta'$ or else $\beta' \leq \beta$, which is absurd. So $\beta$ is unique, and Lemma 1.13 is proved. □

The root $\beta$ in Lemma 1.13 is called the *highest root* of $R$. For each of the lattices of Theorem 1.2, the coefficients $m_i$ of the highest root are depicted in Fig. 1.8.

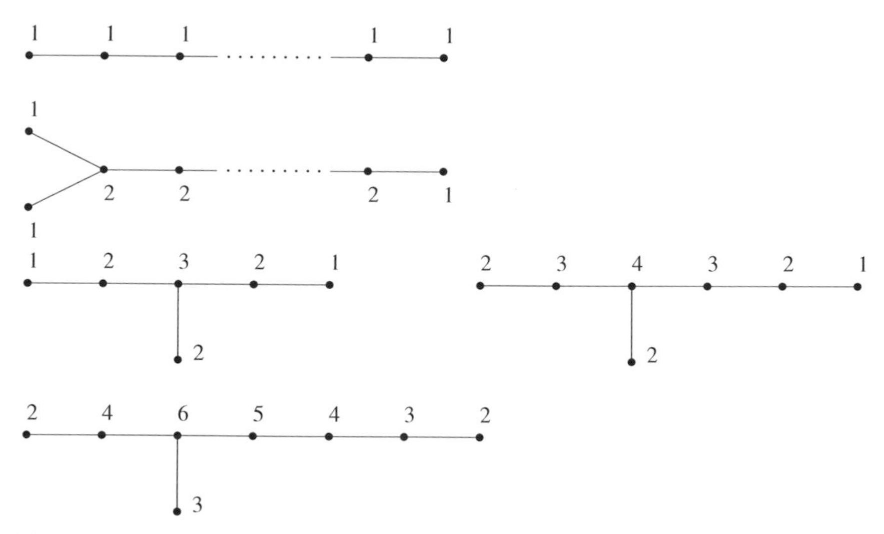


**Fig. 1.8** Coefficients of the highest roots of the irreducible root lattices

Recall that $s_\alpha$ for a root $\alpha$ is the reflection at the hyperplane $\alpha^\perp$.

**Lemma 1.14.** *Let $\alpha = e_j$ for some $1 \leq j \leq n$. Then $s_\alpha$ permutes the positive roots other than $\alpha$.*

*Proof.* Let $\beta \in R^+ - \{\alpha\}$, $\beta = \sum_{i=1}^{n} k_i e_i$, where $k_i \in \mathbb{Z}$, $k_i \geq 0$. The root $\beta$ cannot be proportional to $\alpha$, since otherwise $\beta^2 = 2$ and $k_j \geq 0$ would imply $\beta = \alpha$, contrary to the assumption on $\beta$. Therefore we must have $k_i > 0$ for some $i \neq j$. Now $s_\alpha(\beta) = \beta - (\beta \cdot \alpha)\alpha$, so that the coefficient of $e_i$ in $s_\alpha(\beta)$ is still $k_i$. This means that $s_\alpha(\beta)$ has at least one positive coefficient, and hence $s_\alpha(\beta) \in R^+$. Since $\alpha$ is the image of $-\alpha$, $s_\alpha(\beta) \neq \alpha$. This proves Lemma 1.14. □

**Lemma 1.15.** *Let $\rho = \frac{1}{2} \sum_{\alpha>0} \alpha$. Then $\rho \cdot e_i = 1$ for all $1 \leq i \leq n$.*

*Proof.* By Lemma 1.14, $s_{e_i}(\rho) = \rho - e_i$ and hence $\rho \cdot e_i = 1$ for all $1 \leq i \leq n$. □

The vector $\rho$ is called the *Weyl vector* of $\Gamma$.

**Lemma 1.16.** *Let $\beta = \sum_{i=1}^{n} m_i e_i$ be the highest root. Then*

$$\sum_{i=1}^{n} m_i + 1 = h,$$

*where $h$ is the Coxeter number of $\Gamma$.*

*Proof.* By Lemma 1.15,

$$\rho \cdot \beta = \sum m_i (\rho \cdot e_i) = \sum m_i .$$

On the other hand,

$$\begin{aligned} \rho \cdot \beta &= \frac{1}{2} \sum_{\alpha > 0} \alpha \cdot \beta \\ &= \sum_{\alpha > 0} \frac{(\alpha \cdot \beta)^2}{(\beta \cdot \beta)} - 1 \qquad (\text{since } \alpha \cdot \beta \in \{0,1\} \text{ unless } \alpha = \beta) \\ &= \frac{1}{2} \sum_{\alpha \in R} \frac{(\alpha \cdot \beta)^2}{(\beta \cdot \beta)} - 1 . \end{aligned}$$

Applying Proposition 1.6 we get

$$\sum m_i + 1 = h.$$

This proves Lemma 1.16. □

Let $\beta$ be the highest root of $\Gamma$. Let $e_{n+1} := -\beta$. Then according to Lemma 1.13, $e_{n+1} \cdot e_i \leq 0$ for all $1 \leq i \leq n$. According to Fig. 1.8, $-1 \leq e_{n+1} \cdot e_i$ except for $n = 1$ (type $A_1$), where $e_2 \cdot e_1 = -2$. We represent the inner product $-2$ in the Coxeter-Dynkin diagram by connecting the corresponding vertices by two edges. Then the Coxeter-Dynkin diagram corresponding to $(e_1, \ldots, e_{n+1})$ is called the corresponding *extended Coxeter-Dynkin diagram.* The extended Coxeter-Dynkin diagrams are the diagrams shown in Fig. 1.9.

In contrast to the extended diagrams, we call the Coxeter-Dynkin diagrams of the irreducible root lattices the *ordinary* diagrams.

Let $\rho$ be the Weyl vector. We shall compute $\rho^2$. For that purpose let $(e_1^*, \ldots, e_n^*)$ be the dual basis of $(e_1, \ldots, e_n)$. Then Lemma 1.15 implies that $\rho = e_1^* + \ldots + e_n^*$. So

$$\rho^2 = \sum_{i,j} e_i^* \cdot e_j^* = \sum_{i,j} b_{ij},$$

where $B = ((b_{ij}))$ is the inverse matrix of the matrix $A = ((e_i \cdot e_j))$ (cf. Sect. 1.1).

We shall list the matrices $B$ in each case of Theorem 1.2.

$A_n$: Consider a fundamental system of roots $\{e_1, \ldots, e_n\}$ with Coxeter-Dynkin diagram

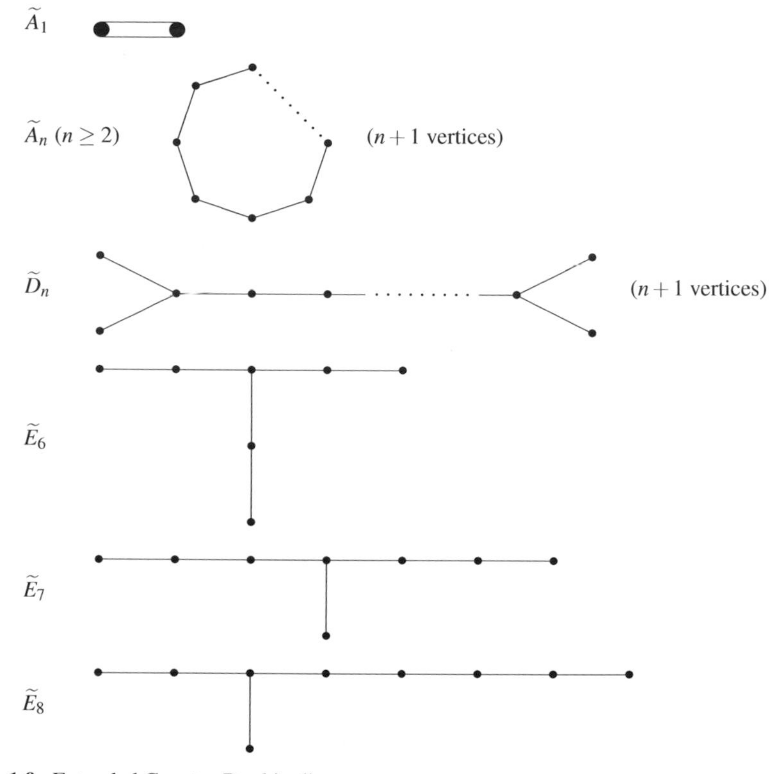


**Fig. 1.9** Extended Coxeter-Dynkin diagrams

Then

$$e_i^* = \frac{1}{n+1}[(n-i+1)e_1 + 2(n-i+1)e_2 + \ldots + (i-1)(n-i+1)e_{i-1} \\ + i(n-i+1)e_i + i(n-i)e_{i+1} + \ldots + ie_n]$$

for $i = 1, \ldots, n$.

$D_n$: Consider a fundamental system of roots $\{e_1, \ldots, e_n\}$ with Coxeter-Dynkin diagram

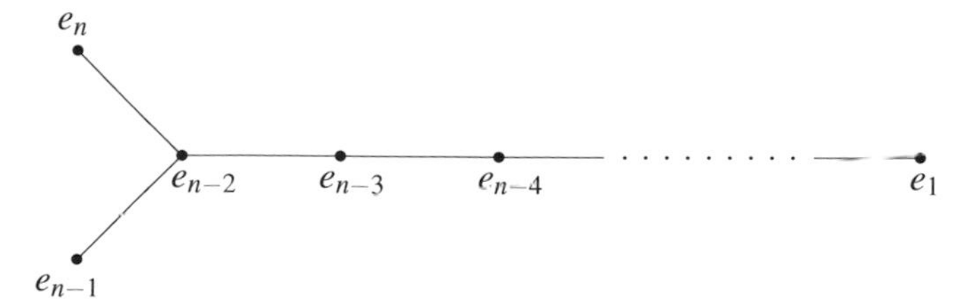

Then

$$\begin{aligned}
e_i^* &= e_1 + 2e_2 + \ldots + (i-1)e_{i-1} + i(e_i + e_{i+1} + \ldots + e_{n-2}) + \frac{1}{2}i(e_{n-1} + e_n) \\
&\quad (1 \le i \le n-2), \\
e_{n-1}^* &= \frac{1}{2}[e_1 + 2e_2 + \ldots + (n-2)e_{n-2} + \frac{1}{2}ne_{n-1} + \frac{1}{2}(n-2)e_n], \\
e_n^* &= \frac{1}{2}[e_1 + 2e_2 + \ldots + (n-2)e_{n-2} + \frac{1}{2}(n-2)e_{n-1} + \frac{1}{2}ne_n].
\end{aligned}$$

$E_8$: Consider a fundamental system of roots $\{e_1, \ldots, e_8\}$ with Coxeter-Dynkin diagram

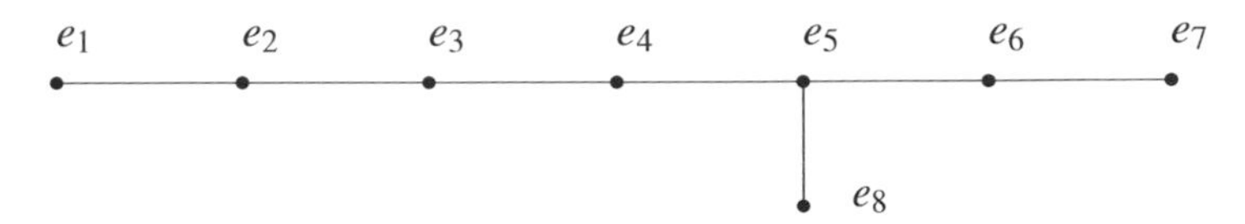


Then we have

$$B = \begin{pmatrix} 2 & 3 & 4 & 5 & 6 & 4 & 2 & 3 \\ 3 & 6 & 8 & 10 & 12 & 8 & 4 & 6 \\ 4 & 8 & 12 & 15 & 18 & 12 & 6 & 9 \\ 5 & 10 & 15 & 20 & 24 & 16 & 8 & 12 \\ 6 & 12 & 18 & 24 & 30 & 20 & 10 & 15 \\ 4 & 8 & 12 & 16 & 20 & 14 & 7 & 10 \\ 2 & 4 & 6 & 8 & 10 & 7 & 4 & 5 \\ 3 & 6 & 9 & 12 & 15 & 10 & 5 & 8 \end{pmatrix}.$$

$E_7$: Corresponding to the set of roots $\{e_2, \ldots, e_8\}$ of $E_8$ we have

$$B = \begin{pmatrix} \frac{3}{2} & 2 & \frac{5}{2} & 3 & 2 & 1 & \frac{3}{2} \\ 2 & 4 & 5 & 6 & 4 & 2 & 3 \\ \frac{5}{2} & 5 & \frac{15}{2} & 9 & 6 & 3 & \frac{9}{2} \\ 3 & 6 & 9 & 12 & 8 & 4 & 6 \\ 2 & 4 & 6 & 8 & 6 & 3 & 4 \\ 1 & 2 & 3 & 4 & 3 & 2 & 2 \\ \frac{3}{2} & 3 & \frac{9}{2} & 6 & 4 & 2 & \frac{7}{2} \end{pmatrix}.$$

$E_6$: Corresponding to the set of roots $\{e_3, \ldots, e_8\}$ of $E_8$ we have

$$B = \begin{pmatrix} \frac{4}{3} & \frac{5}{3} & 2 & \frac{4}{3} & \frac{2}{3} & 1 \\ \frac{5}{3} & \frac{10}{3} & 4 & \frac{8}{3} & \frac{4}{3} & 2 \\ 2 & 4 & 6 & 4 & 2 & 3 \\ \frac{4}{3} & \frac{8}{3} & 4 & \frac{10}{3} & \frac{5}{3} & 2 \\ \frac{2}{3} & \frac{4}{3} & 2 & \frac{5}{3} & \frac{4}{3} & 1 \\ 1 & 2 & 3 & 2 & 1 & 2 \end{pmatrix}.$$

Computing $\sum b_{ij}$ in each case one gets

**Lemma 1.17.** *If $h$ is the Coxeter number of $\Gamma$ then*

$$\rho^2 = \frac{1}{12} nh(h+1).$$

For Sect. 4.3 we need the following lemma.

**Lemma 1.18.** *Let $\Gamma$ be an irreducible root lattice, and let $(e_1, \ldots, e_n)$ be a fundamental system of roots of $\Gamma$. Then the vectors of $\Gamma^*$ that have inner product 0 or 1 with all positive roots of $\Gamma$ (with respect to $(e_1, \ldots, e_n)$) form a complete set of coset representatives for $\Gamma^*/\Gamma$.*

*Proof.* Let $y \in \Gamma^*$, $y \neq 0$, be a vector with $y \cdot \alpha$ equal to 0 or 1 for all positive roots of $\Gamma$. If $\beta$ denotes the highest root of $\Gamma$, then $y \cdot \beta = 1$, because otherwise $y \cdot \beta = 0$ which would imply $y \cdot e_i = 0$ for all $i$, $1 \leq i \leq n$, a contradiction. Since

$$\beta = \sum_{i=1}^{n} m_i e_i \quad , \text{ all } m_i > 0,$$

we must have $y \cdot e_j = 1$ for some $j$ with $m_j = 1$ and $y \cdot e_i = 0$ for all $i \neq j$. This implies $y = e_j^*$, where $e_j^*$ is an element of the dual basis $(e_1^*, \ldots, e_n^*)$ of $(e_1, \ldots, e_n)$ satisfying $e_j^* \cdot e_i = \delta_{ij}$. Let $J$ be the set of indices $i$, $1 \leq i \leq n$, with $m_i = 1$. Examining the above list of matrices $B$ we see that the vectors $e_j^*$, $j \in J$, form a complete set of nonzero coset representatives for $\Gamma^*/\Gamma$. This proves Lemma 1.18. □

# Chapter 2
# Theta Functions and Weight Enumerators

## 2.1 The Theta Function of a Lattice

Let $\Gamma \subset \mathbb{R}^n$ be a lattice. We associate to $\Gamma$ a function which is defined on the upper half plane

$$\mathbb{H} = \{\tau \in \mathbb{C} \mid \operatorname{Im} \tau > 0\} \subset \mathbb{C}.$$

For $\tau \in \mathbb{H}$ let $q = e^{2\pi i \tau}$.

**Definition.** The *theta function* of the lattice $\Gamma$ is the function

$$\vartheta_\Gamma(\tau) := \sum_{x \in \Gamma} q^{\frac{1}{2} x \cdot x}$$

for $\tau \in \mathbb{H}$.

We shall show later that this is a well defined function. Now let $\Gamma \subset \mathbb{R}^n$ be an integral lattice. Then the numbers $x \cdot x$ for $x \in \Gamma$ are nonnegative integers. The coefficient of $q^{\frac{1}{2}\widetilde{r}}$ is equal to the number of $x \in \Gamma$ with $x \cdot x = \widetilde{r}$.

Now let $\Gamma$ be an even integral lattice. Then

$$\vartheta_\Gamma(\tau) = \sum_{r=0}^{\infty} a_r q^r,$$

where $a_r = |\{x \in \Gamma \mid x \cdot x = 2r\}|$, so $\vartheta_\Gamma$ is the generating function for the finite numbers $a_r$. The number $a_r$ counts the points of the lattice $\Gamma$ lying on a sphere of radius $\sqrt{2r}$ around the origin. This shows that $a_r$ grows like the area of the sphere, hence like $\left(\sqrt{2r}\right)^{n-1}$. But

$$\lim \sqrt[r]{\left(\sqrt{2r}\right)^{n-1}} = 1,$$

so the series $\sum_{r=0}^{\infty} a_r q^r$ converges for $|q| < 1$, which is equivalent to $\tau \in \mathbb{H}$.

Our aim will be to prove the following theorem:

**Theorem 2.1.** *Let $\Gamma$ be an even unimodular lattice in $\mathbb{R}^n$. Then*
(i) $n \equiv 0 \pmod 8$.
(ii) $\vartheta_\Gamma$ *is a modular form of weight* $\frac{n}{2}$.

## 2.2 Modular Forms

The group

$$\mathrm{SL}_2(\mathbb{Z}) = \left\{ g = \begin{pmatrix} a & b \\ c & d \end{pmatrix} \middle| \, a,b,c,d \in \mathbb{Z},\ ad-bc=1 \right\}$$

acts on $\mathbb{H}$ by fractional linear transformations

$$\tau \longmapsto g(\tau) = \frac{a\tau+b}{c\tau+d}.$$

Note that $\operatorname{Im} g(\tau) = \frac{\operatorname{Im}\tau}{|c\tau+d|^2} > 0$ if $\operatorname{Im}\tau > 0$.

The center $\{\pm 1\}$ of $\mathrm{SL}_2(\mathbb{Z})$ acts trivially. The quotient $G := \mathrm{SL}_2(\mathbb{Z})/\{\pm 1\}$ is called the *modular group*.

Let $S$ and $T$ be the elements of $G$ represented by the matrices

$$S = \begin{pmatrix} 0 & -1 \\ 1 & 0 \end{pmatrix} \quad \text{and} \quad T = \begin{pmatrix} 1 & 1 \\ 0 & 1 \end{pmatrix}.$$

The elements $S$ and $T$ act on $\mathbb{H}$ as follows:

$$S: \tau \longmapsto -\frac{1}{\tau}, \qquad T: \tau \longmapsto \tau+1.$$

We shall show that $G$ is generated by these elements.

Two points $\tau, \tau' \in \mathbb{H}$ are *equivalent* under $G$ if there is an element $g \in G$ such that $\tau = g(\tau')$. This is clearly an equivalence relation. The equivalence class of an element $\tau \in \mathbb{H}$ is also called the *orbit* of $\tau$ under $G$. A *fundamental domain* for the action of $G$ on $\mathbb{H}$ is a subset $D \subset \mathbb{H}$ such that every element of $\mathbb{H}$ is equivalent to an element in $D$, and two elements of $D$ are equivalent only if they lie on the boundary of $D$. The *stabilizer* of a point $\tau \in \mathbb{H}$ in $G$ is the subgroup $G_\tau = \{g \in G \,|\, g(\tau) = \tau\}$.

Let $D$ be the subset of $\mathbb{H}$ formed of all points $\tau$ such that $|\tau| \geq 1$ and $|\operatorname{Re}\tau| \leq \frac{1}{2}$ (cf. Fig. 2.1). Let $\eta = e^{2\pi i/3}$.

**Theorem 2.2.** (i) *The subset $D$ is a fundamental domain for the action of $G$ on $\mathbb{H}$.*
(ii) *Let $\tau \in D$. Then $G_\tau = \{1\}$ for $\tau \notin \{i, \eta, -\overline{\eta}\}$, $G_i = \{1, S\}$, $G_\eta = \{1, ST, (ST)^2\}$, and $G_{-\overline{\eta}} = \{1, TS, (TS)^2\}$.*
(iii) *The group $G$ is generated by $S$ and $T$.*

*Proof.* (Cf. [81, Chap. VII, §1].) Let $G'$ be the subgroup of $G$ generated by $S$ and $T$.

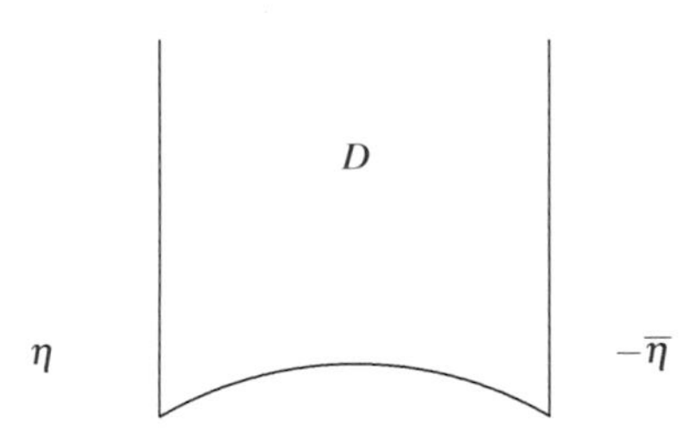


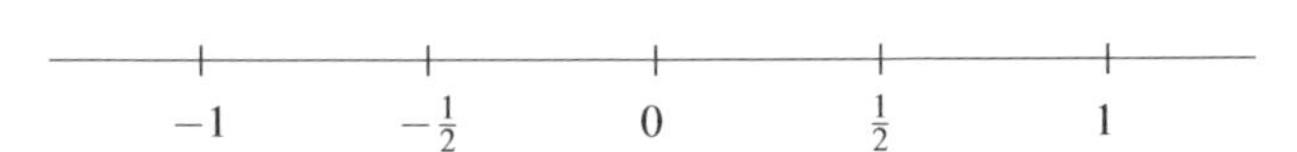


**Fig. 2.1** A fundamental domain for the modular group

a) We shall first show that for every $\tau \in \mathbb{H}$ there exists an element $g \in G'$ such that $g(\tau) \in D$. Let $\tau \in \mathbb{H}$ and choose

$$g = \begin{pmatrix} a & b \\ c & d \end{pmatrix} \in G'$$

such that $-\frac{1}{2} \leq \operatorname{Re} g(\tau) \leq \frac{1}{2}$ and

$$\operatorname{Im} g(\tau) = \frac{\operatorname{Im} \tau}{|c\tau + d|^2}$$

is maximal. This is possible since $\operatorname{Im} g(\tau) \geq \operatorname{Im} \tau$ and thus $|c\tau + d| \leq 1$. But for a fixed $\tau \in \mathbb{H}$ there are only finitely many pairs $(c,d) \in \mathbb{Z} \times \mathbb{Z}$ such that $|c\tau + d| \leq 1$. We claim that $|g(\tau)| \geq 1$ and hence $g(\tau) \in D$. For suppose that $|g(\tau)| < 1$. Then

$$\operatorname{Im} Sg(\tau) = \frac{\operatorname{Im} g(\tau)}{|g(\tau)|^2} > \operatorname{Im} g(\tau),$$

which is impossible by the choice of $g$.

b) We now show that if $\tau, \tau' \in D$, $\tau \neq \tau'$, are equivalent, then $\operatorname{Re} \tau = \frac{1}{2}$, $\operatorname{Re} \tau' = -\frac{1}{2}$ and $\tau' = \tau - 1$, or $\operatorname{Re} \tau = -\frac{1}{2}$, $\operatorname{Re} \tau' = \frac{1}{2}$ and $\tau' = \tau + 1$, or $|\tau| = |\tau'| = 1$ and $\tau' = -\frac{1}{\tau}$. Together with a), this shows that $D$ is a fundamental domain for the action of the group $G$ on $\mathbb{H}$. At the same time we shall prove assertion (ii). Let $\tau \in D$. Let

$$g = \begin{pmatrix} a & b \\ c & d \end{pmatrix} \in G$$

such that $\tau' = g(\tau) \in D$. We may assume that $\operatorname{Im} g(\tau) \geq \operatorname{Im} \tau$ because otherwise we can interchange the points $\tau$ and $\tau'$ and replace $g$ by its inverse. As we have already seen above, this implies $|c\tau+d|^2 \leq 1$. Write $\tau = x+iy$. Since $\tau \in D$, $y \geq \frac{1}{2}\sqrt{3}$. Hence $|c\tau+d|^2 = c^2y^2+(cx+d)^2 \leq 1$ implies $c^2 \leq \frac{4}{3}$. Since $c \in \mathbb{Z}$, $c = \pm 1$ or $c = 0$.

Let us first consider the case $c = 0$. Then $d = 1$ and $\tau' = g(\tau) = \tau + b$. But $\tau' \in D$, $\tau \neq \tau'$, implies $b = 1$ and $\operatorname{Re}\tau = -\frac{1}{2}$, $\operatorname{Re}\tau' = \frac{1}{2}$, or $b = -1$ and $\operatorname{Re}\tau = \frac{1}{2}$, $\operatorname{Re}\tau' = -\frac{1}{2}$.

Now suppose that $c = 1$. Since $|c\tau+d| \leq 1$ we have three possibilities:

($\alpha$) $|\tau| = 1, d = 0,$

($\beta$) $\tau = \eta, d = 1,$

($\gamma$) $\tau = -\overline{\eta}, d = -1.$

In case ($\alpha$), the condition $ad-bc = 1$ implies $b = -1$ and $g(\tau) = a - \frac{1}{\tau}$. If $\tau \notin \{\eta, -\overline{\eta}\}$, then we must have $a = 0$ and hence $g = S$. This also shows that $G_i = \{1, S\}$. If $\tau = \eta$ then either $a = 0$ and $g = S$ or $a = -1$ and

$$g = \begin{pmatrix} -1 & -1 \\ 1 & 0 \end{pmatrix} = T^{-1}S = (ST)^{-1} = (ST)^2.$$

This also shows that $(ST)^2 \in G_\eta$. Finally, if $\tau = -\overline{\eta}$, then either $a = 0$ and $g = S$ or $a = 1$ and

$$g = \begin{pmatrix} 1 & -1 \\ 1 & 0 \end{pmatrix} = TS \in G_{-\overline{\eta}}.$$

In case ($\beta$), $b = a-1$ and $g(\eta) = a - \frac{1}{\eta+1}$. This implies $a = 0$ and

$$g = \begin{pmatrix} 0 & -1 \\ 1 & 1 \end{pmatrix} = ST \in G_\eta.$$

The case ($\gamma$) can be treated similarly.

Finally the case $c = -1$ can be reduced to the case $c = 1$ by changing the signs of $a$, $b$, $c$, and $d$.

c) We finally show that $G = G'$. Then assertion (iii) follows. Let $g \in G$ and let $\tau_0$ be a point in the interior of $D$. By a) there is an element $g' \in G'$ with $g'g(\tau_0) \in D$. By b) we must have $g'g = 1$. Hence $g \in G'$. This completes the proof of Theorem 2.2. □

**Definition.** Let $k$ be an even positive integer. A holomorphic function $f : \mathbb{H} \longmapsto \mathbb{C}$ is called a *modular form of weight k*, if the following conditions are satisfied :

(i) $f\left(\frac{a\tau+b}{c\tau+d}\right) = (c\tau+d)^k f(\tau)$ for all $\begin{pmatrix} a & b \\ c & d \end{pmatrix} \in \mathrm{SL}_2(\mathbb{Z})$,

(ii) $f$ has a power series expansion in $q = e^{2\pi i\tau}$, i.e., $f$ is holomorphic at infinity $\tau = i\infty$.

Note that the first condition implies that $f(\tau) = f(\tau+b)$ for all $b \in \mathbb{Z}$, so $f$ is periodic. This implies that $f$ has a Laurent series expansion in $q = e^{2\pi i\tau}$:

$$f(\tau) = \sum_{-\infty}^{+\infty} a_r q^r.$$

The second condition means that this is actually a power series, i.e., $a_r = 0$ for $r < 0$.

In view of Theorem 2.2, a modular form $f$ of weight $k$ is given by a series

$$f(\tau) = \sum_{r=0}^{\infty} a_r q^r$$

which converges for $|q| < 1$ (i.e., $\tau \in \mathbb{H}$ ), and which satisfies the identity

$$f\left(-\frac{1}{\tau}\right) = \tau^k f(\tau).$$

So in order to show that $\vartheta_\Gamma$ is a modular form of weight $\frac{n}{2}$, we have to show that $\frac{n}{2}$ is even, and that $\vartheta_\Gamma$ satisfies the identity

$$\vartheta_\Gamma\left(-\frac{1}{\tau}\right) = \tau^{\frac{n}{2}} \vartheta_\Gamma(\tau).$$

This will follow from the Poisson summation formula.

## 2.3 The Poisson Summation Formula

Let $\Gamma \subset \mathbb{R}^n$ be an arbitrary lattice, and let $f : \mathbb{R}^n \to \mathbb{C}$ be a function which satisfies the following conditions (V1), (V2), and (V3):

(V1) $\int_{\mathbb{R}^n} |f(x)|\, dx < \infty$

(V2) The series $\sum_{x \in \Gamma} |f(x+u)|$ converges uniformly for all $u$ belonging to a compact subset of $\mathbb{R}^n$.

The condition (V1) implies the existence of the *Fourier transform* $\widehat{f}$ of $f$ , which is defined by the formula

$$\widehat{f}(y) := \int_{\mathbb{R}^n} f(x) e^{-2\pi i x \cdot y}\, dx.$$

The condition (V2) implies that the function $F(u) := \sum_{x \in \Gamma} f(x+u)$ is continuous on $\mathbb{R}^n$.

The third condition will be:

(V3) The series $\sum_{y \in \Gamma^*} \widehat{f}(y)$ is absolutely convergent.

**Theorem 2.3 (Poisson summation formula).** *Let $f : \mathbb{R}^n \longmapsto \mathbb{C}$ be a function satisfying the conditions (V1), (V2), and (V3). Then*

$$\sum_{x \in \Gamma} f(x) = \frac{1}{\operatorname{vol}(\mathbb{R}^n/\Gamma)} \sum_{y \in \Gamma^*} \widehat{f}(y).$$

*Proof.* We assume first that $\Gamma = \mathbb{Z}^n$. The function

$$F(u) := \sum_{x \in \Gamma} f(x+u)$$

is continuous by (V2) and periodic in $u$, i.e., $F(u+x) = F(u)$ for all $x \in \mathbb{Z}^n$. Hence it can be developed into a Fourier series

$$\sum_{y \in \mathbb{Z}^n} e^{2\pi i u \cdot y} a(y),$$

where $a(y) := \int_{[0,1]^n} F(t) e^{-2\pi i y \cdot t}\, dt$. We shall show that

$$a(y) = \widehat{f}(y). \tag{2.1}$$

Then condition (V3) implies that the Fourier series of $F$ converges absolutely and uniformly, hence it converges to a continuous function, hence to $F$. Then

$$F(0) = \sum_{x \in \Gamma} f(x) = \sum_{y \in \mathbb{Z}^n} \widehat{f}(y),$$

which is the Poisson summation formula. The identity (2.1) follows from the following equalities:

$$\begin{aligned} a(y) &= \int_{[0,1]^n} \sum_{x \in \mathbb{Z}^n} f(x+t) e^{-2\pi i t \cdot y}\, dt \\ &= \sum_{x \in \mathbb{Z}^n} \int_{[0,1]^n} f(x+t) e^{-2\pi i (t+x) \cdot y}\, dt \\ &= \sum_{x \in \mathbb{Z}^n} \int_{x+[0,1]^n} f(t') e^{-2\pi i t' \cdot y}\, dt' \\ &= \widehat{f}(y). \end{aligned}$$

In the general case $\Gamma = M \cdot \mathbb{Z}^n$, where $M \in \mathrm{GL}_n(\mathbb{Z})$. (The matrix $M$ was denoted by $C$ in Sect. 1.1.) From Sect. 1.1 one can derive that $\Gamma^* = (M^t)^{-1} \cdot \mathbb{Z}^n$. Now

$$\sum_{x \in \Gamma} f(x) = \sum_{x \in \mathbb{Z}^n} f(Mx) = \sum_{x \in \mathbb{Z}^n} f_M(x) = \sum_{y \in \mathbb{Z}^n} \widehat{f_M}(y),$$

where

$$\begin{aligned} \widehat{f_M}(y) &= \int_{\mathbb{R}^n} f(Mt) e^{-2\pi i t \cdot y}\, dt \\ &= \frac{1}{\det M} \int_{\mathbb{R}^n} f(t') e^{-2\pi i (M^{-1} t') \cdot y}\, dt' \qquad (t = M^{-1} t') \\ &= \frac{1}{\mathrm{vol}\,(\mathbb{R}^n / \Gamma)} \widehat{f}\left((M^t)^{-1} y\right) \qquad (M^{-1} t' \cdot y = t' \cdot (M^t)^{-1} y). \end{aligned}$$

But since $\Gamma^* = (M^t)^{-1} \cdot \mathbb{Z}^n$, it follows that

$$\sum_{x\in\Gamma} f(x) = \sum_{y\in\mathbb{Z}^n} \widehat{f_M}(y) = \frac{1}{\mathrm{vol}\,(\mathbb{R}^n/\Gamma)} \sum_{y\in\Gamma^*} \widehat{f}(y).$$

This proves the Poisson summation formula. □

**Lemma 2.1.** *The function* $f : x \longmapsto e^{-\pi x^2}$ $(x \in \mathbb{R}^n)$ *is equal to its Fourier transform, i.e.,*

$$\int_{\mathbb{R}^n} e^{-\pi x^2} e^{-2\pi i x\cdot y}\, dx = e^{-\pi y^2}.$$

*Proof.* By Fubini's theorem it suffices to show the formula for $n = 1$. Let $\widehat{f}(y) = \int_{\mathbb{R}} e^{-\pi x^2} e^{-2\pi i xy}\, dx$ for $y \in \mathbb{R}$ be the Fourier transform of $f(x)$. Integration by parts with $g(x) = e^{2\pi i xy}$, $h'(x) = -2\pi x e^{-\pi x^2}$ shows that

$$\begin{aligned} \widehat{f}'(y) &= \int_{\mathbb{R}} -2\pi i x e^{-\pi x^2} e^{-2\pi i xy}\, dx \\ &= -2\pi y \widehat{f}(y). \end{aligned}$$

From this equality it follows that differentiating the quotient

$$\frac{\widehat{f}(y)}{e^{-\pi y^2}}$$

yields 0. Hence there is a constant $c$ such that $\widehat{f}(y) = c e^{-\pi y^2}$. But

$$\widehat{f}(0) = \int_{\mathbb{R}} e^{-\pi x^2}\, dx = 1.$$

Hence $c = 1$, which proves the lemma. □

## 2.4 Theta Functions as Modular Forms

We want to prove Theorem 2.1. In order to be able to apply the results of Sect. 2.3, we first show the following lemma.

**Lemma 2.2.** *Let* $\Gamma \subset \mathbb{R}^n$ *be a lattice. Then the series*

$$\sum_{x\in\Gamma} q^{\frac{1}{2}x\cdot x} = \sum_{x\in\Gamma} e^{\pi i \tau x^2}$$

*converges uniformly absolutely for all* $\tau$ *with* $\mathrm{Im}\,\tau \geq v_0 > 0$.

*Proof.* Let $\Gamma = M \cdot \mathbb{Z}^n$ for $M \in \mathrm{GL}_n(\mathbb{R})$, and let

$$\varepsilon := \min_{|x|=1} (Mx)^2.$$

Then $\varepsilon > 0$ and $(Mx)^2 \geq \varepsilon x^2$ for all $x \in \mathbb{R}^n$. This yields the following estimation

$$\sum_{x\in\Gamma} \left| e^{\pi i \tau x^2} \right| = \sum_{x\in\mathbb{Z}^n} \left| e^{\pi i \tau (Mx)^2} \right| \leq \sum_{x\in\mathbb{Z}^n} e^{-\pi v_0 \varepsilon x^2} = \left( \sum_{r=-\infty}^{\infty} e^{-\pi v_0 \varepsilon r^2} \right)^n < \infty.$$

This proves the lemma. □

Lemma 2.2 shows that $\vartheta_\Gamma$ is well defined and holomorphic for $\tau \in \mathbb{H}$.

**Proposition 2.1.** *We have the identity*

$$\vartheta_\Gamma \left( -\frac{1}{\tau} \right) = \left( \frac{\tau}{i} \right)^{\frac{n}{2}} \frac{1}{\mathrm{vol}\,(\mathbb{R}^n/\Gamma)} \vartheta_{\Gamma^*}(\tau).$$

*Proof.* Since both sides of the identity are holomorphic in $\tau \in \mathbb{H}$, it suffices to prove this formula when $\tau = it$ with $t \in \mathbb{R}$, $t > 0$. The Fourier transform of $e^{-\pi(\frac{1}{t})x^2}$ is $\left(\sqrt{t}\right)^n e^{-\pi t y^2}$. This follows from Lemma 2.1 with $\widetilde{x} = \frac{1}{\sqrt{t}} x$. Therefore the Poisson summation formula yields

$$\begin{aligned} \vartheta_\Gamma \left( -\frac{1}{it} \right) &= \sum_{x\in\Gamma} e^{\pi i (-\frac{1}{it}) x^2} = \sum_{x\in\Gamma} e^{-\pi(\frac{1}{t})x^2} \\ &= \frac{1}{\mathrm{vol}\,(\mathbb{R}^n/\Gamma)} \sum_{y\in\Gamma^*} \left(\sqrt{t}\right)^n e^{-\pi t y^2} \\ &= t^{\frac{n}{2}} \frac{1}{\mathrm{vol}\,(\mathbb{R}^n/\Gamma)} \vartheta_{\Gamma^*}(it), \end{aligned}$$

which proves Proposition 2.1. □

*Proof of Theorem 2.1.* Let $\Gamma$ be an even unimodular lattice in $\mathbb{R}^n$. We first prove (i). Suppose that $n$ is not divisible by 8. We may also assume that $n \equiv 4 \pmod 8$ because we can replace $\Gamma$, if necessary, by $\Gamma \perp \Gamma$ or $\Gamma \perp \Gamma \perp \Gamma \perp \Gamma$. Proposition 2.1 then yields

$$\vartheta_\Gamma \left( -\frac{1}{\tau} \right) = (-1)^{\frac{n}{4}} \tau^{\frac{n}{2}} \vartheta_\Gamma(\tau) = -\tau^{\frac{n}{2}} \vartheta_\Gamma(\tau),$$

since $\Gamma^* = \Gamma$ and $\mathrm{vol}\,(\mathbb{R}^n/\Gamma) = 1$. Since $\vartheta_\Gamma$ is invariant under $T$,

$$\vartheta_\Gamma\left((TS)\tau\right) = -\tau^{\frac{n}{2}} \vartheta_\Gamma(\tau).$$

From this formula one derives

$$\vartheta_\Gamma\left((TS)^3\tau\right) = -(-1)^{\frac{n}{2}} \vartheta_\Gamma(\tau) = -\vartheta_\Gamma(\tau).$$

But $(TS)^3 = 1$, which is a contradiction. This proves (i).

To prove (ii) it remains to show the identity at the end of Sect. 2.2:

$$\vartheta_\Gamma\left(-\frac{1}{\tau}\right) = \tau^{\frac{n}{2}}\vartheta_\Gamma(\tau).$$

But this follows from Proposition 2.1, since $n \equiv 0 \pmod 8$, $\Gamma = \Gamma^*$ and $\mathrm{vol}(\mathbb{R}^n/\Gamma) = 1$. This shows that $\vartheta_\Gamma$ is a modular form of weight $\frac{n}{2}$, and finishes the proof of Theorem 2.1. □

## 2.5 The Eisenstein Series

We now consider some further very important examples of modular forms:

**Definition.** Let $k \in \mathbb{Z}$, $k$ even, $k > 2$. The series

$$G_k(\tau) = \sum_{\substack{(m,n)\in\mathbb{Z}^2\\(m,n)\neq(0,0)}} \frac{1}{(m\tau+n)^k}$$

is called the *Eisenstein series of index k.*

**Proposition 2.2.** *The Eisenstein series $G_k(\tau)$ ($k$ even, $k > 2$) is a modular form of weight $k$.*

It is clear from the definition that $G_k(\tau)$ is invariant under $T : \tau \longmapsto \tau + 1$ and that

$$G_k\left(-\frac{1}{\tau}\right) = \tau^k G_k(\tau).$$

We have to prove that $G_k : \mathbb{H} \longrightarrow \mathbb{C}$ is holomorphic and that $G_k$ has a power series expansion in $q = e^{2\pi i\tau}$. We need the following lemma.

**Lemma 2.3.** *Let $L$ be a lattice in $\mathbb{C} \cong \mathbb{R}^2$. Then the series*

$$\sum_{\substack{\gamma\in L\\ \gamma\neq 0}} \frac{1}{|\gamma|^\sigma}$$

*converges for $\sigma > 2$.*

*Proof.* Let $(\omega_1, \omega_2)$, $\omega_1, \omega_2 \in \mathbb{C}$, be a basis for $L$. Let $P_m$ be the parallelogram with the vertices $\pm m\omega_1 \pm m\omega_2$ for $m \in \mathbb{Z}$, $m \geq 1$ (this means the boundary of the convex hull of these points ). Then (cf. Fig. 2.2)

$$|P_m \cap L| = 8m.$$

Let $r = \min_{\tau\in P_1} |\tau|$. Then $mr = \min_{\tau\in P_m} |\tau|$. Then

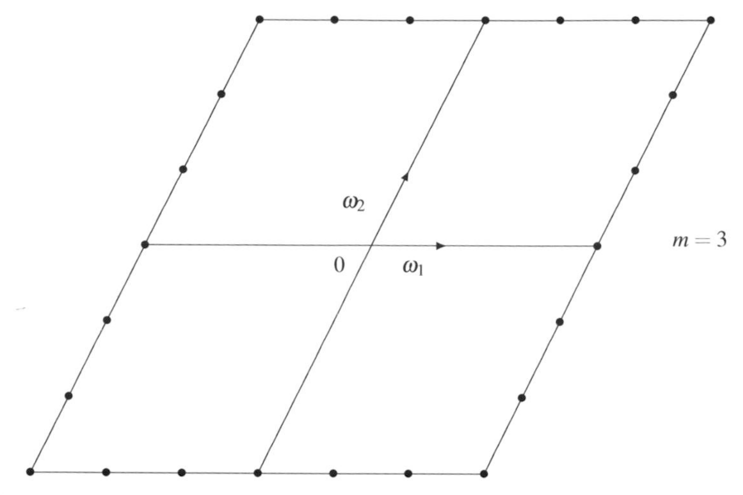


**Fig. 2.2** The parallelogram $P_m$

$$\sum_{\gamma \in P_m \cap L} \frac{1}{|\gamma|^\sigma} \leq 8m \frac{1}{(mr)^\sigma}$$

Hence the series

$$\sum_{\substack{\gamma \in L \\ \gamma \neq 0}} \frac{1}{|\gamma|^\sigma} = \sum_{m=1}^{\infty} \sum_{\gamma \in P_m \cap L} \frac{1}{|\gamma|^\sigma}$$

is majorized by the series

$$\sum_{m=1}^{\infty} (8m) \frac{1}{(mr)^\sigma} = \text{const.} \sum_{m=1}^{\infty} \frac{1}{m^{\sigma-1}}.$$

But this series converges for $\sigma > 2$, since it can be majorized by the integral

$$\int_1^{\infty} \frac{dx}{x^{\sigma-1}}.$$

□

**Lemma 2.4.** *$G_k$ is holomorphic in the upper half plane $\mathbb{H}$.*

*Proof of Lemma 2.4.* First suppose that $\tau$ is contained in the fundamental domain $D$. Then

$$|m\tau + n|^2 = m^2 \tau\overline{\tau} + 2mn \operatorname{Re} \tau + n^2 \geq m^2 - mn + n^2 = |m\eta + n|^2$$

This implies

$$|m\tau + n|^{-k} \leq |m\eta + n|^{-k}$$

for $k$ even, $k \geq 2$. By Lemma 2.3, the series

$$\sum_{\substack{(m,n)\in\mathbb{Z}^2\\(m,n)\neq(0,0)}} \frac{1}{|m\eta+n|^k}$$

converges for $k>2$. This shows that the series $G_k(\tau)$ converges uniformly on every compact subset of $D$. Applying the result to $G_k(g^{-1}\tau)$ for $g\in \mathrm{SL}_2(\mathbb{Z})/\{\pm 1\}$ shows that the same holds for each of the transforms $gD$ of $D$, which cover $\mathbb{H}$. Thus $G_k$ is the limit of a sequence of holomorphic functions, which converges uniformly on every compact subset of $\mathbb{H}$. Therefore $G_k$ is holomorphic on $\mathbb{H}$. □

We now want to derive the power series expansion of $G_k$ in $q=e^{2\pi i\tau}$. First consider the *Riemann $\zeta$-function*

$$\zeta(s) := \sum_{m=1}^{\infty}\frac{1}{m^s}.$$

The right hand series converges for $s\in\mathbb{C}$, $\operatorname{Re} s>1$, since for $s=\sigma+it$, $\sigma,t\in\mathbb{R}$,

$$|m^s| = \left|e^{s\log m}\right| = e^{\sigma\log m} = m^{\sigma}.$$

So the Riemann $\zeta$-function $\zeta$ is defined and holomorphic for $s\in\mathbb{C}$ with $\operatorname{Re} s>1$.

**Proposition 2.3.** *Let $\sigma_l(r) := \sum_{d|r} d^l$ denote the sum of the $l$-th powers of the positive divisors of $r$. Then one has*

$$G_k(\tau) = 2\zeta(k) + \frac{2(2\pi i)^k}{(k-1)!}\sum_{r=1}^{\infty}\sigma_{k-1}(r)q^r.$$

*Proof.* We consider the function $\tau\longmapsto \pi\operatorname{cotg}\pi\tau$. Since

$$\pi\operatorname{cotg}\pi\tau = \pi i\frac{e^{i\pi\tau}+e^{-i\pi\tau}}{e^{i\pi\tau}-e^{-i\pi\tau}},$$

$\tau\longmapsto\pi\operatorname{cotg}\pi\tau$ is a meromorphic function for $\tau\in\mathbb{C}$, which is holomorphic for $\tau\in\mathbb{H}$ and has poles on the real axis, namely for $\tau\in\mathbb{Z}$. Since

$$\lim_{\tau\to 0}\pi\tau\operatorname{cotg}\pi\tau = 1,$$

the residue in $\tau=0$ is equal to 1, and the same is true for all $\tau\in\mathbb{Z}$. By the theorem of Mittag-Leffler (see for example [1, Chap. 5, §2.1] )

$$\pi\operatorname{cotg}\pi\tau = \frac{1}{\tau} + \sum_{m=1}^{\infty}\left(\frac{1}{\tau-m}+\frac{1}{\tau+m}\right).$$

On the other hand

$$\pi\operatorname{cotg}\pi\tau = \pi i\frac{e^{2\pi i\tau}+1}{e^{2\pi i\tau}-1} = \pi i\frac{q+1}{q-1}$$

$$= \pi i \left(1 - 2 \cdot \frac{1}{1-q}\right) = \pi i \left(1 - 2 \sum_{n=0}^{\infty} q^n\right).$$

Combining these two formulas, we get

$$\frac{1}{\tau} + \sum_{m=1}^{\infty} \left(\frac{1}{\tau - m} + \frac{1}{\tau + m}\right) = \pi i \left(1 - 2 \sum_{n=0}^{\infty} q^n\right).$$

Differentiating this formula $(k-1)$-times yields the identity (valid for $k \geq 2$):

$$(-1)^k \sum_{m=-\infty}^{\infty} \frac{(k-1)!}{(\tau+m)^k} = (2\pi i)^k \sum_{n=0}^{\infty} n^{k-1} q^n.$$

Now (for $k$ even)

$$\begin{aligned} G_k(\tau) &= \sum_{\substack{(m,n) \in \mathbb{Z}^2 \\ (m,n) \neq (0,0)}} \frac{1}{(n\tau + m)^k} \\ &= 2\zeta(k) + 2 \sum_{n=1}^{\infty} \sum_{m \in \mathbb{Z}} \frac{1}{(n\tau + m)^k}. \end{aligned}$$

Applying the above identity with $\tau$ replaced by $n\tau$, we get ($k$ even!)

$$\begin{aligned} G_k(\tau) &= 2\zeta(k) + \frac{2(2\pi i)^k}{(k-1)!} \sum_{d=1}^{\infty} \sum_{n=0}^{\infty} n^{k-1} q^{nd} \\ &= 2\zeta(k) + \frac{2(2\pi i)^k}{(k-1)!} \sum_{r=1}^{\infty} \sigma_{k-1}(r) q^r. \end{aligned}$$

This proves Proposition 2.3. □

Proposition 2.2 follows from Lemma 2.4 and Proposition 2.3.

One normalizes the Eisenstein series to get the constant term $a_0 = 1$ in the power series expansion in $q$.

**Definition.** The *normalized Eisenstein series* is the series

$$E_k(\tau) := \frac{1}{2\zeta(k)} G_k(\tau).$$

Now the values $\zeta(k)$ for even $k$ can be expressed in terms of the *Bernoulli numbers* $B_k$. The Bernoulli numbers $B_k$ are defined by

$$\frac{x}{e^x - 1} = \sum_{k=0}^{\infty} B_k \frac{x^k}{k!}.$$

In particular $B_0 = 1$, $B_1 = -\frac{1}{2}$, $B_k = 0$ for $k > 1, k$ odd, and, for example, $B_2 = \frac{1}{6}$, $B_4 = -\frac{1}{30}$, $B_6 = \frac{1}{42}$.

**Proposition 2.4.** *If $k$ is an even integer, $k \geq 2$, then*

$$\zeta(k) = -\frac{(2\pi i)^k}{2 \cdot k!} B_k.$$

*Proof.* We consider the function $\tau \longmapsto \pi\tau \operatorname{cotg} \pi\tau$. Then putting $x = 2\pi i\tau$ in the definition of $B_k$ yields

$$\begin{aligned} \pi\tau \operatorname{cotg} \pi\tau &= \pi i\tau \frac{e^{\pi i\tau} + e^{-\pi i\tau}}{e^{\pi i\tau} - e^{-\pi i\tau}} = \pi i\tau \frac{e^{2\pi i\tau} + 1}{e^{2\pi i\tau} - 1} \\ &= \pi i\tau + \frac{2\pi i\tau}{e^{2\pi i\tau} - 1} = 1 + \sum_{k=2}^{\infty} B_k \frac{(2\pi i\tau)^k}{k!} \\ &= 1 + \sum_{k=2}^{\infty} B_k \frac{(2\pi i)^k}{k!} \tau^k. \end{aligned}$$

On the other hand, by the formula in the proof of Proposition 2.3, we have

$$\begin{aligned} \pi\tau \operatorname{cotg} \pi\tau &= 1 + \sum_{m=1}^{\infty} \left( \frac{\tau}{\tau + m} + \frac{\tau}{\tau - m} \right) \\ &= 1 + 2 \sum_{m=1}^{\infty} \frac{\tau^2}{\tau^2 - m^2} = 1 - 2 \sum_{m=1}^{\infty} \frac{\frac{\tau^2}{m^2}}{1 - \frac{\tau^2}{m^2}} \\ &= 1 - 2 \sum_{m=1}^{\infty} \sum_{l=1}^{\infty} \frac{\tau^{2l}}{m^{2l}}. \end{aligned}$$

Comparing coefficients yields the formula of Proposition 2.4. □

**Corollary 2.1.** *With the Bernoulli numbers $B_k$ one has*

$$E_k(\tau) = 1 - \frac{2k}{B_k} \sum_{r=1}^{\infty} \sigma_{k-1}(r) q^r.$$

*Proof.* By Proposition 2.3, the definition of $E_k$, and Proposition 2.4, the coefficient in front of the infinite sum is equal to

$$\frac{2(2\pi i)^k}{(k-1)! 2\zeta(k)} = -\frac{(2\pi i)^k}{(k-1)!} \cdot \frac{2k!}{(2\pi i)^k B_k} = -\frac{2k}{B_k}.$$

This proves the corollary. □

So, for example,

$$E_4(\tau) = 1 + 240 \sum_{r=1}^{\infty} \sigma_3(r) q^r.$$

$$E_6(\tau) = 1 - 504\sum_{r=1}^{\infty} \sigma_5(r)q^r.$$

Now consider

$$E_4^3 - E_6^2.$$

This is a modular form of weight 12 with constant term $a_0 = 0$ in the power series expansion

$$\sum_{r=0}^{\infty} a_r q^r$$

in $q$. A modular form with constant term $a_0 = 0$ in the power series expansion in $q$ is called a *cusp form*. So $E_4^3 - E_6^2$ is a nontrivial cusp form whose power series expansion in $q$ starts with

$$(3\cdot 240 + 2\cdot 504)q = 1728q.$$

We shall compute the coefficient of $q^2$ as an exercise. We have

$$\begin{aligned} E_4(\tau) &= 1 + 240q + 240\cdot 9q^2 + \cdots, \\ E_6(\tau) &= 1 - 504q - 504\cdot 33q^2 + \cdots. \end{aligned}$$

So the coefficient of $q^2$ in the power series expansion of $E_4^3 - E_6^2$ is

$$3\cdot 240\cdot 9 + 3\cdot 240\cdot 240 + 2\cdot 504\cdot 33 - 504\cdot 504 = -1728\cdot 24.$$

So

$$\left(E_4^3 - E_6^2\right)(\tau) = 0 + 1728q + 1728(-24)\cdot q^2 + \cdots.$$

In fact one has the following identity which we state without proof (for a proof see e.g. [81, pp. 95–96]):

$$\Delta := \frac{1}{1728}(E_4^3 - E_6^2) = q\prod_{r=1}^{\infty}(1-q^r)^{24}.$$

## 2.6 The Algebra of Modular Forms

Let $k$ be an integer. The modular forms of weight $k$ form a $\mathbb{C}$-vector space which we denote by $M_k$. We have $M_k = 0$ for $k$ odd (by definition or extending the definition of a modular form to allow odd weights). In this section we want to determine the vector spaces $M_k$ for $k$ even.

We shall first derive a formula about the orders of the zeros of a modular form. Let $f$ be a holomorphic function on $\mathbb{H}$, not identically zero, and let $p$ be a point of $\mathbb{H}$. Let

$$f(\tau) = \sum_{r=0}^{\infty} a_r(\tau - p)^r$$

be the power series expansion of $f$ in $p$. The smallest integer $r$ with $a_r \neq 0$ is called the *order* of $f$ at $p$ and is denoted by $\nu_p(f)$.

Let $G = \mathrm{SL}_2(\mathbb{Z})/\{\pm 1\}$ be the modular group. When $f$ is a modular form of weight $k$, the identity

$$f(\tau) = (c\tau + d)^{-k} f\left(\frac{a\tau + b}{c\tau + d}\right)$$

shows that $\nu_{g(p)}(f) = \nu_p(f)$ for all $g \in G$. In other terms, $\nu_p(f)$ depends only on the image of $p$ in $\mathbb{H}/G$. We define $\nu_{i\infty}(f)$ to be the smallest integer $r$ with $a_r \neq 0$ in the power series expansion

$$f(\tau) = \sum_{r=0}^{\infty} a_r q^r$$

of $f$ in $q = e^{2\pi i \tau}$.

We denote by $e_p$ the order of the stabilizer $G_p := \{g \in G \mid g(p) = p\}$ of $G$ of the point $p$. We have $e_p = 2$ if $p$ is congruent modulo $G$ to $i$, $e_p = 3$ if $p$ is congruent to $\eta$ modulo $G$, and $e_p = 1$ otherwise.

**Theorem 2.4.** *Let $f$ be a modular form of weight $k$, which is not identically zero. One has*

$$\nu_{i\infty}(f) + \sum_{p \in \mathbb{H}/G} \frac{1}{e_p} \nu_p(f) = \nu_{i\infty}(f) + \frac{1}{2}\nu_i(f) + \frac{1}{3}\nu_\eta(f) + \sum_{\substack{p \in \mathbb{H}/G \\ p \neq [i],[\eta]}} \nu_p(f) = \frac{k}{12},$$

*where $[i], [\eta]$ denote the classes of $i, \eta$ respectively in $\mathbb{H}/G$.*

One can define a complex analytic structure on the compactification $\widehat{\mathbb{H}/G} := \mathbb{H}/G \cup \{i\infty\}$ of $\mathbb{H}/G$. This turns $\widehat{\mathbb{H}/G}$ into a Riemann surface. One can show that this Riemann surface is isomorphic to the Riemann sphere $\mathbb{P}_1(\mathbb{C}) = S^2 = \mathbb{C} \cup \{i\infty\}$. Theorem 2.4 can then be derived from the *residue theorem* on $\mathbb{P}_1(\mathbb{C})$. The dimensions of the spaces $M_k$ of modular forms of weight $k$ can then be computed using the *Riemann-Roch theorem* on $\mathbb{P}_1(\mathbb{C})$. A proof following these lines can be found in [28].

We give an elementary proof using the residue theorem in the complex plane $\mathbb{C}$ (cf. [81, Chap. VII, §3]). We first have to verify that the sum in Theorem 2.4 makes sense, i.e., that $f$ has only a finite number of zeros in $\mathbb{H}$ modulo $G$. Indeed, since $f$ is holomorphic at $i\infty$, there exists $\rho > 0$ such that $f$ has no zeros for $0 < |q| < \rho$; this means that $f$ has no zeros outside the compact subset $D_\rho$ of the fundamental domain $D$ defined by the inequality $\operatorname{Im} \tau \leq e^{2\pi\rho}$. Since $f$ is holomorphic in $\mathbb{H}$, it has only a finite number of zeros in $D_\rho$.

*Proof of Theorem 2.4.* We integrate the logarithmic derivative of $f$ along the boundary of the compact subset $D_\rho$ of the fundamental domain $D$ of the action of $G$ on $\mathbb{H}$, cf. Fig. 2.3. Let $\mathfrak{C}$ be this contour, oriented counterclockwise. By the choice of $\rho$, $D_\rho$ contains a representative of each zero of $f$ in $\mathbb{H}$.

We first suppose that there are no zeros of $f$ on the boundary of $D_\rho$. By the argument principle we have

$$\frac{1}{2\pi i} \int_{\mathfrak{C}} \frac{f'(\tau)}{f(\tau)} d\tau = \sum_{p \in \mathbb{H}/G} \nu_p(f).$$

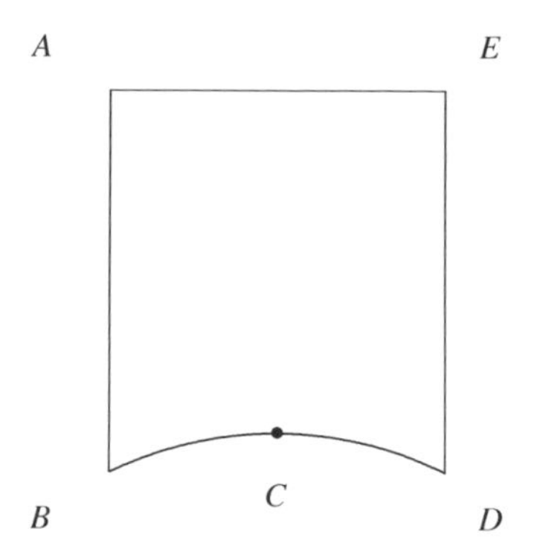


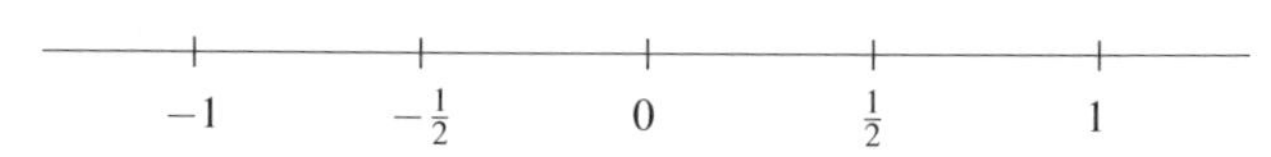


**Fig. 2.3** Contour $\mathfrak{C}$

On the other hand

$$\int_{\mathfrak{C}} \frac{f'(\tau)}{f(\tau)} d\tau =$$
$$\int_A^B \frac{f'(\tau)}{f(\tau)} d\tau + \int_B^C \frac{f'(\tau)}{f(\tau)} d\tau + \int_C^D \frac{f'(\tau)}{f(\tau)} d\tau + \int_D^E \frac{f'(\tau)}{f(\tau)} d\tau + \int_E^A \frac{f'(\tau)}{f(\tau)} d\tau.$$

Now the first and fourth integral on the right-hand side cancel each other out, since the transformation $T \in G$ transforms the arc $AB$ into the arc $ED$ and $f(T\tau) = f(\tau)$. The transformation $S \in G$ transforms the arc $BC$ into the arc $DC$. Since $f$ is a modular form of weight $k$, $f(S\tau) = \tau^k f(\tau)$. This implies that

$$\frac{f'(S\tau)}{f(S\tau)} = k \cdot \frac{1}{\tau} + \frac{f'(\tau)}{f(\tau)}.$$

Hence we get

$$\begin{aligned} \int_B^C \frac{f'(\tau)}{f(\tau)} d\tau + \int_C^D \frac{f'(\tau)}{f(\tau)} d\tau &= \int_B^C \left( \frac{f'(\tau)}{f(\tau)} - \frac{f'(S\tau)}{f(S\tau)} \right) d\tau \\ &= -\int_B^C \frac{k}{\tau} d\tau \\ &= 2\pi i \frac{k}{12}. \end{aligned}$$

There remains to compute the integral

$$\int_E^A \frac{f'(\tau)}{f(\tau)} d\tau.$$

Under the change of variables $q = e^{2\pi i\tau}$, the arc $EA$ is transformed into the unit circle centred around the origin, oriented clockwise. By the choice of $\mathfrak{C}$, the origin is the only possible zero of $f$ inside this circle. Therefore we get

$$\int_E^A \frac{f'(\tau)}{f(\tau)} d\tau = -2\pi i v_{i\infty}(f).$$

Putting everything together yields the formula of Theorem 2.4 in the case when there are no zeros of $f$ on the boundary of $D_\rho$.

If there are zeros of $f$ on the boundary of $D_\rho$, we modify the contour $\mathfrak{C}$ by taking small arcs around these zeros as indicated in Fig. 2.4. We compute the limit of the integral of

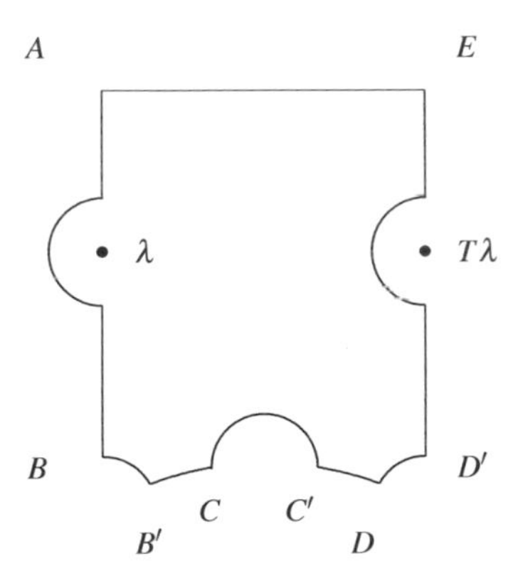


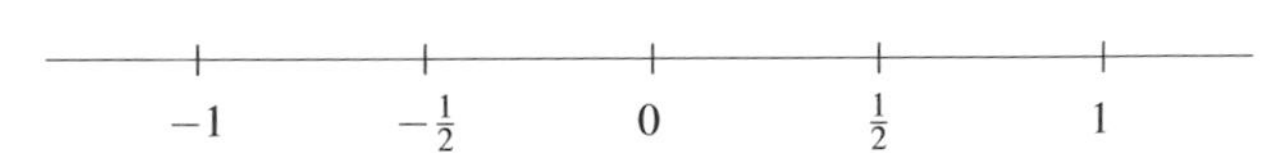


**Fig. 2.4** Modified contour $\mathfrak{C}$

$f'/f$ over $\mathfrak{C}$ as the radii of these circles tend to 0.

Let $\omega$ be the circle centred around $\eta$ which contains the arc $BB'$ and is oriented clockwise. Then

$$\int_\omega \frac{f'(\tau)}{f(\tau)} d\tau = -2\pi i v_\eta(f).$$

When the radius of $\omega$ tends to 0, the angle of the arc $BB'$ tends to $\frac{\pi}{3}$. Therefore the limit of the integral

$$\int_B^{B'} \frac{f'(\tau)}{f(\tau)} d\tau$$

as the radius of $\omega$ tends to 0 is equal to

$$-\frac{2\pi i}{6} v_\eta(f).$$

We get the same contribution by the limit of the integral over the arc $DD'$ which lies on a small circle around $-\overline{\eta}$.

Similarly we obtain for the limit of the integral of $f'/f$ over the arc $CC'$

$$-\frac{2\pi i}{2} v_i(f).$$

Finally, let $\lambda$ be a zero of $f$ on the boundary of $D$ which is different from $\eta$, $-\overline{\eta}$, and $i$. Then the orbit of $\lambda$ under $G$ intersects the boundary of $D$ in two points, and the contributions from the two small semi-circles around these points cancel each other. This completes the proof of Theorem 2.4. □

Let $M_k^0$ denote the $\mathbb{C}$-vector space of cusp forms of weight $k$. By definition, $M_k^0$ is the kernel of the linear form $f \longmapsto f(i\infty)$ on $M_k$, which associates to $f$ the coefficient $a_0 = f(i\infty)$ of the power series expansion

$$f(\tau) = \sum_{r=0}^{\infty} a_r q^r$$

of $f$ in $q$. Thus we have $\dim M_k/M_k^0 \leq 1$. Moreover, for $k \geq 4$, $k$ even, the (normalized) Eisenstein series $E_k$ is an element of $M_k$ with $E_k(i\infty) = 1 \neq 0$. Hence we have

$$M_k = M_k^0 \oplus \mathbb{C} \cdot E_k \quad (\text{for } k \geq 4,\ k \text{ even}).$$

**Theorem 2.5.** (i) *We have $M_k = 0$ for $k$ odd, for $k < 0$, and for $k = 2$.*
(ii) *We have $M_0 = \mathbb{C}$, $M_0^0 = 0$, and, for $k = 4, 6, 8, 10$, $M_k^0 = 0$, $M_k = \mathbb{C} \cdot E_k$.*
(iii) *Multiplication by $\Delta = \frac{1}{1728}(E_4^3 - E_6^2)$ defines an isomorphism of $M_{k-12}$ onto $M_k^0$.*

*Proof.* Let $f$ be a nonzero element of $M_k$ and consider the formula of Theorem 2.4

$$v_{i\infty}(f) + \frac{1}{2} v_i(f) + \frac{1}{3} v_\eta(f) + \sum_{\substack{p \in \mathbb{H}/G \\ p \neq [i],[\eta]}} v_p(f) = \frac{k}{12}.$$

Since all the terms on the left hand side of this formula are $\geq 0$, we have $k \geq 0$. The formula also shows that the case $k = 2$ is impossible. This proves (i).

To prove (iii), let $f = \Delta$, $k = 12$. Then the right-hand side of the above formula is equal to 1. We have already seen in Sect. 2.5 that $v_{i\infty}(\Delta) = 1$. The above formula then shows that $\Delta$ does not vanish on $\mathbb{H}$. Therefore multiplication by $\Delta$ is injective.

To prove surjectivity, let $f$ be an element of $M_k^0$, and set $g = \frac{f}{\Delta}$. Then $g$ is of weight $k-12$. One has

$$\nu_p(g) = \nu_p(f) - \nu_p(\Delta) = \begin{cases} \nu_p(f) & \text{if } p \neq i\infty, \\ \nu_p(f) - 1 & \text{if } p = i\infty. \end{cases}$$

By definition, $\nu_{i\infty}(f) \geq 1$. Therefore $\nu_p(g) \geq 0$ for all $p$, and hence $g$ is holomorphic in $\mathbb{H}$ and at infinity. This shows that $g \in M_{k-12}$. This proves (iii).

Finally, we prove (ii). If $k \leq 10$, then we have $k - 12 < 0$. From (i) and (iii) we deduce that $M_k^0 = 0$ and $\dim M_k \leq 1$ for $k \leq 10$. Now (ii) follows, since $E_k$ is a nonzero element of $M_k$.

This completes the proof of Theorem 2.5. □

If we apply the formula of Theorem 2.4 to $f = E_4$ and $k = 4$, then we see that the only zero of $E_4$ in $D$ is $\eta$, and it is of order 1. Similarly we find that the only zero of $E_6$ in $D$ is $i$, and $\nu_i(E_6) = 1$.

Let $M = \sum_{k=0}^{\infty} M_k$ be the sum of the vector spaces $M_k$. This sum is direct, since by the transformation formula, a non-trivial modular form has a unique weight. We have already made use of the fact that multiplication of functions defines a mapping

$$M_k \times M_l \longrightarrow M_{k+l}.$$

This turns $M = \bigoplus_{k=0}^{\infty} M_k$ into a graded algebra.

Let $f_1, \ldots, f_m \in M$. Recall that $f_1, \ldots, f_m$ are called *algebraically dependent*, if there is a non-trivial polynomial $P \in \mathbb{C}[x_1, \ldots, x_m]$ such that $P(f_1, \ldots, f_m)$ is identically zero. Otherwise, $f_1, \ldots, f_m$ are called *algebraically independent*.

**Corollary 2.2.** *The algebra $M$ of modular forms is isomorphic to the polynomial algebra $\mathbb{C}[E_4, E_6]$ of complex polynomials in the Eisenstein series $E_4$ and $E_6$, i.e.,*

$$M = \mathbb{C}[E_4, E_6].$$

*Proof.* (Cf. also [53, pp.10–12].) We first prove that $E_4$ and $E_6$ generate the algebra $M$. For this purpose, we have to show that each vector space $M_k$ is generated by monomials in $E_4$ and $E_6$. This is clear for $k \leq 6$ by (i) and (ii) of Theorem 2.5. Let $k$ be an even integer with $k \geq 8$. Let $f \in M_k$. Let $g = E_4^{\alpha} E_6^{\beta}$, where $\alpha = r$, $\beta = 0$, if $k = 4r$, and $\alpha = r - 1$, $\beta = 1$, if $k = 4r + 2$. Then $g$ is a modular form of weight $k$ which does not vanish at infinity. Therefore we can find $\lambda \in \mathbb{C}$ such that $f - \lambda g$ is a cusp form. If $k \leq 10$, then $f - \lambda g = 0$ by Theorem 2.5(ii) and we are done. If $k \geq 12$, by Theorem 2.5(iii) there exists $h \in M_{k-12}$ such that

$$f - \lambda g = \Delta h.$$

Now the proof is finished by induction on $k$.

In order to prove that $M$ is isomorphic to the polynomial algebra $\mathbb{C}[E_4, E_6]$, it remains to prove that $E_4$ and $E_6$ are algebraically independent. Assume that $P$ is a non-trivial complex polynomial such that $P(E_4, E_6) \equiv 0$. Then every monomial of weighted degree $k$ occurring in $P$ with a nonzero coefficient is a modular form of weight $k$. Since $M$ is the direct sum of the vector spaces $M_k$, all such weights have to be the same. Therefore we can assume that $P$ is a weighted homogeneous polynomial of weighted degree $k$. Moreover, $E_4$ divides each monomial of $P$. Otherwise, we would have

$$P(E_4, E_6) = cE_6^{\alpha} + E_4\widetilde{P}(E_4, E_6),$$

where $c \in \mathbb{C}$, $c \neq 0$, $\alpha$ is a non-negative integer with $\alpha \geq 1$, and $\widetilde{P}$ a non-trivial weighted homogeneous polynomial of weighted degree $k-4$. Since $E_4(\eta) = 0$ and $E_6(\eta) \neq 0$, this is impossible. Therefore $c = 0$, and we find a relation $\widetilde{P}(E_4, E_6) \equiv 0$ of lower degree. Using induction, we derive a contradiction. This concludes the proof of Corollary 2.2. □

As a first application of this corollary we show the uniqueness of the $E_8$-lattice.

**Proposition 2.5.** *Let $\Gamma$ be an even unimodular lattice in $\mathbb{R}^8$. Then $\Gamma$ is isomorphic to $E_8$.*

*Proof.* Let $\vartheta_\Gamma$ be the theta function of $\Gamma$. By Theorem 2.1, $\vartheta_\Gamma$ is a modular form of weight 4. By Theorem 2.5, $M_4 = \mathbb{C} \cdot E_4$. Since $\vartheta_\Gamma$ has constant term 1 in the power series expansion in $q$, $\vartheta_\Gamma = E_4$. So

$$\vartheta_\Gamma(\tau) = 1 + 240q + \text{higher order terms}.$$

This implies that there are 240 roots in $\Gamma$. These roots generate a (possibly reducible) root sublattice of rank $\leq 8$. This lattice must be of type $E_8$, since any other root lattice of rank $\leq 8$ contains less roots (cf. Table 1.1). Therefore $\Gamma$ is isomorphic to $E_8$. □

**Remark 2.1** The history of the $E_8$-lattice starts in the second half of the last century. In 1867, H. J. S. Smith [86] proved the existence of an even unimodular 8-dimensional lattice in a non-constructive way. Later, explicit constructions were given by A. Korkine and G. Zolotareff in 1873 [49, 50, 51], and by H. Minkowski (1884) [63]. The $E_8$-lattice occurs in many branches of mathematics. For example Milnor used it in his construction of exotic spheres (~1958). He used a basis with Coxeter-Dynkin diagram (see [62])

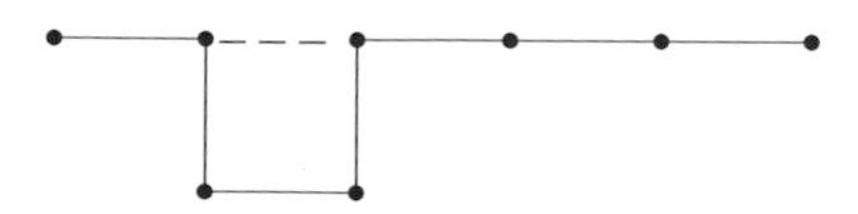

where $e_i \bullet - - - \bullet e_j$ means $e_i \cdot e_j = 1$, i.e., $\angle(e_i, e_j) = 60°$. In fact one has to interchange straight and dashed lines in order to get Milnor's original basis.

**Exercise 2.1** Show that this diagram defines the $E_8$-lattice, i.e., reconstruct the standard diagram of $E_8$.

## 2.7 The Weight Enumerator of a Code

Let $C \subset \mathbb{F}_q^n$ be a (linear) code of type $[n,k,d]$. Recall that we denote by $w(u)$ the weight of a codeword $u \in C$ (cf. Sect. 1.2).

**Definition.** The *Hamming weight enumerator* of $C$ is the polynomial

$$\begin{aligned} W_C(X,Y) &:= \sum_{u \in C} X^{n-w(u)} Y^{w(u)} \\ &= \sum_{i=0}^{n} A_i X^{n-i} Y^i, \end{aligned}$$

where $A_i$ is the number of codewords of weight $i$. This is a homogeneous polynomial of degree equal to the length of the code. Note that

$$\sum_{i=0}^{n} A_i = q^k.$$

**Example 2.1** For the Hamming code $H \subset \mathbb{F}_2^7$

$$W_H(X,Y) = X^7 + 7X^4Y^3 + 7X^3Y^4 + Y^7,$$

and for the extended Hamming code $\widetilde{H} \subset \mathbb{F}_2^8$

$$W_{\widetilde{H}}(X,Y) = X^8 + 14X^4Y^4 + Y^8.$$

We now want to give applications of the results on theta functions and modular forms.

**Proposition 2.6.** *Let $C \subset \mathbb{F}_2^n$ be a self-dual doubly even code. Then $n \equiv 0 \pmod 8$.*

*Proof.* This follows from Theorem 2.1, since the lattice $\Gamma_C$ is an even unimodular lattice in $\mathbb{R}^n$ by Proposition 1.3. There is also an independent proof, using only results from coding theory. □

**Proposition 2.7.** *Let $n = 24$. Let $C \subset \mathbb{F}_2^{24}$ be a self-dual doubly even code, and let $W_C(X,Y)$ be the Hamming weight enumerator of $C$. Then $A_i = 0$ for $i \not\equiv 0 \pmod 4$ and*

$$A_8 = 759 - 4A_4.$$

*Proof.* We denote by $\vartheta_C$ the theta function $\vartheta_{\Gamma_C}$ of the lattice $\Gamma_C$. Then

$$\vartheta_C(\tau) = \sum_{r=0}^{\infty} a_r q^r,$$

where $a_1 (a_2)$ is the number of elements $x = \frac{1}{\sqrt{2}}(c + 2y) \in \Gamma_C$, $c \in C$, $y \in \mathbb{Z}^{24}$, with $x^2 = 2$ (resp. $4$). So

$$a_1 = 24 \cdot 2 + 16A_4,$$

since $a_1$ is the number of pairs $(c,y)$, $c \in C$, $y \in \mathbb{Z}^{24}$ with $c^2+4cy+4y^2=4$, and $4=(\pm1)^2+(\pm1)^2+(\pm1)^2+(\pm1)^2=(\pm2)^2$ are the only ways to write 4 as a sum of squares. Similarly

$$a_2 = 2^8 A_8 + 16A_4 \cdot 20 \cdot 2 + \binom{24}{2} \cdot 4.$$

So we can derive a relation between $A_4$ and $A_8$ from a relation between $a_1$ and $a_2$. □

**Lemma 2.5.** *Let $f$ be a modular form of weight 12 and*

$$f(\tau) = \sum_{r=0}^{\infty} a_r q^r$$

*its power series expansion in $q$. Then*

$$a_2 = -24a_1 + 196560a_0.$$

*Proof.* By Corollary 2.2 we have only to verify this for $E_4^3$ and $E_6^2$, or, more conveniently, for $\Delta = \frac{1}{1728}(E_4^3 - E_6^2)$ and $E_6^2$. Now

$$\Delta = \frac{1}{1728}(E_4^3 - E_6^2) = 0 \cdot 1 + q - 24q^2 + \ldots,$$

as we have seen at the end of Sect. 2.5, and

$$\begin{aligned} E_6 &= 1 - 504q - 504 \cdot 33q^2 + \ldots, \\ E_6^2 &= 1 - 2 \cdot 504q + (504^2 - 2 \cdot 504 \cdot 33)q^2 + \ldots \\ &= 1 - 1008q + 220752q^2 + \ldots, \end{aligned}$$

but $220752 = 196560 + 24 \cdot 1008$. This proves the lemma. □

**Proposition 2.8.** *Let $\Gamma \subset \mathbb{R}^{24}$ be an even unimodular lattice. Then*

$$a_2 = 196560 - 24a_1.$$

*Proof.* This follows from Lemma 2.5, since the theta function $\vartheta_\Gamma$ is a modular form of weight 12 (Theorem 2.1), and $a_0 = 1$ for $\vartheta_\Gamma$. □

Now Proposition 2.7 follows from Proposition 2.8.

**Example 2.2** Let us consider the code $C = \widetilde{H} \oplus \widetilde{H} \oplus \widetilde{H} \subset \mathbb{F}_2^{24}$. Then

$$\begin{aligned} W_C(X,Y) &= \left(X^8 + 14X^4Y^4 + Y^8\right)^3 \\ &= X^{24} + 42X^{20}Y^4 + (3 \cdot 14^2 + 3)X^{16}Y^8 + \ldots \\ &= X^{24} + 42X^{20}Y^4 + 591X^{16}Y^8 + \ldots, \end{aligned}$$

and $591 = 759 - 168$, which is the assertion of Proposition 2.7.

Propositions 2.7 and 2.8 give rise to the following questions: Is there a linear code $\widetilde{G}$ with $A_4 = 0$? Is there a lattice $\Gamma$ with $a_1 = 0$? Note that such a lattice cannot be of the form $\Gamma_C$, since $a_1 \geq 48$ for $\Gamma_C$.

Let us suppose that there exists a linear code $\widetilde{G}$ with $A_4 = 0$. Then this code determines 759 8-subsets (i.e., subsets with 8 elements) of the set $\{1,2,\ldots,24\}$. Since $\widetilde{G}$ is doubly even, each 5-subset is contained in at most one of these 8-subsets (otherwise add the two codewords; you get a codeword of weight not divisible by 4 or of weight 4 which is excluded by $A_4 = 0$). There are

$$\binom{24}{5} = 759 \cdot \binom{8}{5}$$

5-subsets. Therefore each 5-subset is contained in *exactly one* of the 759 8-subsets. This means that the codewords of $\widetilde{G}$ of weight 8 form a *Steiner system* $S(5,8,24)$. By a theorem of Witt [93], such a system is unique to within a permutation of the set $\{1,2,3,\ldots,24\}$. The group of permutations of this set fixing the Steiner system $S(5,8,24)$ is the Mathieu group $M_{24}$ (see [66, Chap. 20]). This is a sporadic simple group.

**Remark 2.2** The lines in $\mathbb{P}_2(\mathbb{F}_2)$ form a Steiner system $S(2,3,7)$, and this corresponds to the binary $[7,4,3]$ Hamming code.

## 2.8 The Golay Code and the Leech Lattice

We now show that there exists a unique doubly even linear code $\widetilde{G} \subset \mathbb{F}_2^{24}$ with $A_4 = 0$. This is the *extended binary Golay code*, discovered by M. J. E. Golay in 1949 [27]. The presentation in this section is strongly influenced by [7] and by personal communication and private notes by J. H. van Lint.

We shall first show that there is at most one such code. For the proof of this fact we follow [7]. We need some preparations.

**Definition.** Let $S$ be a set with $v$ elements and let $\mathfrak{B}$ be a collection of $k$-subsets of $S$ (which we call *blocks*) with the property that any $t$-subset of $S$ is contained in exactly $\lambda$ blocks. Then the pair $(S,\mathfrak{B})$ is called a *t-design*, more precisely, a *$t$-$(v,k,\lambda)$ design*. The elements of $S$ are called the *points* of the design. A *Steiner system* is a $t$-design with $\lambda = 1$. A $t$-$(v,k,1)$ design is also called a Steiner system $S(t,k,v)$.

For the proof of the following proposition we need some elementary facts from design theory (cf. e.g. [8]). Consider a $t$-$(v,k,\lambda)$ design. Let $S_j$ be a given set of $j$ points with $0 \leq j \leq t$ and denote by $\lambda_j$ the number of blocks containing $S_j$. There are two ways to choose $t$ points starting with the given $j$ ones. One can first choose a block containing $S_j$ and then choose $t-j$ further points from this block. There are

$$\lambda_j \binom{k-j}{t-j}$$

possibilities for this choice. On the other hand one can first choose $t-j$ arbitrary further points and then a block containing all $t$ points. This yields

$$\binom{v-j}{t-j}\lambda$$

choices. So we obtain the equality

$$\lambda_j\binom{k-j}{t-j}=\binom{v-j}{t-j}\lambda. \quad (2.2)$$

It follows that the number $\lambda_j$ is independent of the choice of the original set $S_j$. The number $\lambda_0$ is the total number of blocks and is usually denoted by $b$. The number $\lambda_1$ is the number of blocks containing a given point and is also denoted by $r$.

**Lemma 2.6.** *In a* 2-$(v,k,\lambda)$ *design with* $b=v$ *and* $k=r$, *any two blocks have exactly* $\lambda$ *common points.*

*Proof.* We arrange the characteristic vectors of the blocks as the rows of a $b\times v$ matrix $M$ with entries 0 or 1. This matrix is also called the *incidence matrix* of the design. By our assumptions $b=v$. The conditions that any block contains $k$ points and that any point lies in $r$ blocks can be expressed in terms of $M$ in view of $k=r$ as follows:

$$MJ=kJ=rJ=JM.$$

Here $J$ is the $v\times v$ matrix with all entries equal to 1. The condition that any pair of points lies in $\lambda$ blocks can be expressed as follows:

$$M^tM=(r-\lambda)I+\lambda J,$$

where $I$ is the $v\times v$ identity matrix. Since $M$ commutes with $J$ and $M^t=((r-\lambda)I+\lambda J)M^{-1}$, $M$ also commutes with $M^t$ and we have

$$MM^t=(r-\lambda)I+\lambda J.$$

This means that any two blocks have exactly $\lambda$ points in common. □

**Proposition 2.9.** *There is (essentially) only one* 2-(11,5,2)*-design.*

*Proof.* The collection of blocks of a 2-(11,5,2) design is a set of 5-subsets of an 11-set such that every pair of points is contained in exactly 2 blocks. By Equation 2.2 we have $b=v=11$ and $r=k=5$. Therefore the conditions of Lemma 2.6 are satisfied and it follows from this lemma that any two blocks have 2 common points. Therefore we can conclude as follows. Without loss of generality we can assume that the characteristic vector of the first block is (1 1 1 1 1 0 0 0 0 0 0). The remaining blocks correspond to the 2-subsets of the first five points. One can order these lexicographically. It is an elementary exercise to show that we are left with two possibilities (up to permutation of the points) for the incidence matrix. It is either the matrix

$$\begin{pmatrix}
1&1&1&1&1&0&0&0&0&0&0\\
1&1&0&0&0&1&1&1&0&0&0\\
1&0&1&0&0&1&0&0&1&1&0\\
1&0&0&1&0&0&1&0&1&0&1\\
1&0&0&0&1&0&0&1&0&1&1\\
0&1&1&0&0&0&1&0&0&1&1\\
0&1&0&1&0&0&0&1&1&1&0\\
0&1&0&0&1&1&0&0&1&0&1\\
0&0&1&1&0&1&0&1&0&0&1\\
0&0&1&0&1&0&1&1&1&0&0\\
0&0&0&1&1&1&1&0&0&1&0
\end{pmatrix}$$

or its transpose. But the transpose matrix is obtained by interchanging columns 4 and 5, 7 and 8, and 9 and 10 and the corresponding rows. In this way one sees that up to renumbering of the points one obtains exactly one such design. This proves Proposition 2.9. □

**Definition.** A code $C \subset \mathbb{F}_2^n$ with minimum distance $d = 2e+1$ is called a *perfect code* if one of the following equivalent conditions holds:
(i) Every $x \in \mathbb{F}_2^n$ has distance $\leq e$ to exactly one codeword.
(ii) $|C|(1+\binom{n}{1}+\ldots+\binom{n}{e}) = 2^n$.

**Theorem 2.6.** *Let $C$ be a binary $(24, 2^{12}, 8)$-code containing 0. Then $C$ is a doubly even self-dual linear $[24, 12, 8]$-code, and there is up to equivalence at most one such code.*

*Proof.* Let $C$ be such a code. Take *any* coordinate position and delete it (one calls this process *puncturing* of the code). One finds a $(23, 2^{12}, 7)$-code $C_0$. Since

$$2^{12}(1+\binom{23}{1}+\binom{23}{2}+\binom{23}{3}) = 2^{23},$$

the punctured code $C_0$ is perfect. It follows from this fact that the code $C_0$ has weight enumerator coefficients $A_0 = A_{23} = 1$, $A_7 = A_{16} = 253$, $A_8 = A_{15} = 506$, $A_{11} = A_{12} = 1288$. (This can be seen as follows: Since $C_0$ is perfect, the sphere around $0 \in C_0$ of radius 3 covers the vectors of weight at most 3. The $\binom{23}{4}$ vectors of weight 4 must be covered by spheres of radius 3 around codewords of weight 7, so that $A_7 = \binom{23}{4}/\binom{7}{4} = 253$; next $A_8 = [\binom{23}{5} - A_7\binom{7}{5}]/\binom{8}{5} = 506$; etc.) Now if $C$ contains a word of weight $w$ not divisible by 4, then by suitably puncturing we would find a $C_0$ containing a word of weight $w$ or $w-1$ not equal to 0 or $-1 \pmod 4$, which is a contradiction since *any* puncturing yields the above weight distribution. Hence $C$ has weight enumerator coefficients $A_0 = A_{24} = 1$, $A_8 = A_{16} = 759$, $A_{12} = 2576$. Giving an arbitrary vector in $C$ the role of 0 we see that all distances between codewords are divisible by 4. If $u, v \in C$ then

$$d(u,v) = w(u) + w(v) - 2\langle u, v\rangle,$$

where $\langle\ ,\ \rangle$ is the inner product with values in the real numbers. So $\langle u, v\rangle$ is even, and it follows that $C \subset C^\perp$. But $C^\perp$ is a linear subspace of dimension $24 - \dim\langle C\rangle \leq 12$. It follows that the span $\langle C \rangle$ of $C$ has $\leq 2^{12}$ elements. Since $C$ has $2^{12}$ elements, we see that $C = C^{\perp}$ and $C$ is linear.

Consider a vector $u$ of weight 12 in $C$ such that $u + \overline{u} = \mathbf{1}$ ($\mathbf{1}$: the all one vector) for some vector $\overline{u}$ of weight 12. Consider the code $C_u$ obtained from $C$ by omitting all coordinate positions where $u$ has a 1. This code has word length 12, dimension 11, and all words have even weight. Hence it must be the *even weight code* of length 12 consisting of all vectors in $\mathbb{F}_2^{12}$ of even weight. This means that $C$ has a generator matrix of the form

$$\left(\begin{array}{c|c|c|c} 1^{11} & 1 & 0 & 0^{11} \\ \hline A & (0^{11})^t & (1^{11})^t & I_{11} \end{array}\right),$$

where $I_{11}$ is the $11 \times 11$ unit matrix and $a^k$ is the notation for a row vector with $k$ components of type $a$. The matrix $A$ has the following properties

(i) Every row has weight $\geq 6$.
(ii) Any two rows have distance $\geq 6$.

The first row of the above generator matrix of $C$ corresponds to $u$. Since every row of this matrix has distance $\geq 8$ to $u$, every row has weight 6 in $A$. From equality in (i) one can easily deduce equality in (ii). Let $J$ be the $11 \times 11$ matrix with all entries equal to 1. Then $J - A$ is the incidence matrix of a 2-(11,5,2) design. The uniqueness of $C$ thereby follows from Proposition 2.9. $\square$

We now show the existence of such a code. The construction which we give is due to J. H. Conway (see also [21, Chap. 11]).

We start by constructing a special [6,3,4]-code over $\mathbb{F}_4$. Here $\mathbb{F}_4 = \{0, 1, \omega, \overline{\omega}\}$, $\omega^2 = \omega + 1 = \overline{\omega}$. The words of this code are of the form

$$(a, b, c, f(1), f(\omega), f(\overline{\omega})), \qquad \text{where } f(x) := ax^2 + bx + c.$$

This code is called the *hexacode*. A generator matrix for this code is

$$\begin{pmatrix} 1 & 0 & 0 & 1 & \overline{\omega} & \omega \\ 0 & 1 & 0 & 1 & \omega & \overline{\omega} \\ 0 & 0 & 1 & 1 & 1 & 1 \end{pmatrix}.$$

**Exercise 2.2** Show that the minimum distance of this code is 4, and that no codeword has weight 5.

We now construct a *binary* linear code $\widetilde{G}$ of length 24 as follows. The words of this code are represented by $4 \times 6$ matrices with entries 0 or 1 satisfying the following two rules:

(A) The six column sums and the first row sum have the same parity.
(B) If $r_i$ denotes the $i$-th row ($1 \leq i \leq 4$), then $r_2 + \omega r_3 + \overline{\omega} r_4$ is in the hexacode.

Such a codeword is found in the following way:

(i) Pick a word in the hexacode.
(ii) Choose a parity of the first row.
(iii) Choose the first 5 columns in such a way that the sum in (B) yields the corresponding coordinate in the chosen hexacodeword. The following columns yield the indicated elements of $\mathbb{F}_4$:

$$\begin{matrix} 0 & 0 & 0 & 0 & 1 & 1 & 1 & 1 \\ \begin{pmatrix}0\\0\\0\\0\end{pmatrix} & \begin{pmatrix}1\\1\\1\\1\end{pmatrix} & \begin{pmatrix}1\\0\\0\\0\end{pmatrix} & \begin{pmatrix}0\\1\\1\\1\end{pmatrix} & \begin{pmatrix}1\\1\\0\\0\end{pmatrix} & \begin{pmatrix}0\\0\\1\\1\end{pmatrix} & \begin{pmatrix}0\\1\\0\\0\end{pmatrix} & \begin{pmatrix}1\\0\\1\\1\end{pmatrix} \end{matrix}$$

$$\begin{matrix} \omega & \omega & \omega & \omega & \overline{\omega} & \overline{\omega} & \overline{\omega} & \overline{\omega} \\ \begin{pmatrix}1\\0\\1\\0\end{pmatrix} & \begin{pmatrix}0\\1\\0\\1\end{pmatrix} & \begin{pmatrix}0\\0\\1\\0\end{pmatrix} & \begin{pmatrix}1\\1\\0\\1\end{pmatrix} & \begin{pmatrix}1\\0\\0\\1\end{pmatrix} & \begin{pmatrix}0\\1\\1\\0\end{pmatrix} & \begin{pmatrix}0\\0\\0\\1\end{pmatrix} & \begin{pmatrix}1\\1\\1\\0\end{pmatrix} \end{matrix}$$

(iv) In column 6 we have only one choice.

We observe that rules (A) and (B) are linear. So $\widetilde{G}$ is a linear code. There are $4^3$ choices for (i), 2 choices for (ii), and 2 choices for each of the five columns in (iii). Therefore $\widetilde{G}$ has dimension $2 \cdot 3 + 1 + 5 = 12$. If in (ii) above we take even and in (i) a nonzero hexacodeword, we find a word of $\widetilde{G}$ of weight $\geq 4 \cdot 2 = 8$. If in (ii) above we take even and in (i) the hexacodeword 0, we find either 0 or a word of $\widetilde{G}$ of weight $\geq 2 \cdot 4 = 8$. If in (ii) above we take odd then we clearly find a word of weight $\geq 6$. Weight exactly 6 would imply that the hexacodeword has weight 5, which is not possible by Exercise 2.2. So we find again a word of weight $\geq 8$. Therefore $\widetilde{G}$ has minimum weight $\geq 8$.

We conclude that $\widetilde{G}$ is a [24,12]-code with minimum distance $\geq 8$. So we have shown the existence of a code with the parameters of Theorem 2.6. So there is a unique (up to equivalence) [24,12,8]-code, and this code is called the *extended binary Golay code*.

**Exercise 2.3** Consider those words in $\widetilde{G}$ which have the property that the third and the fifth column of the $4 \times 6$ matrix are equal to the first column, and the fourth and sixth column are equal to the second column. Show that the $4 \times 2$ submatrices of these words consisting of the first and second column form the extended binary $[8,4,4]$ Hamming code.

The (original) *binary Golay code* $G$ is a $[23,12,7]$-code which is obtained from $\widetilde{G}$ by deleting some column of the generator matrix of $\widetilde{G}$. We shall show that it doesn't matter which column to delete. Hence the binary $[23,12,7]$-Golay code is also unique.

In order to show the uniqueness of the Golay code, we give another construction of the extended binary Golay code.

Consider the icosahedron as a graph. Let $A$ be the adjacency matrix of this graph. This is obtained as follows: number the 12 vertices of the icosahedron by $v_1, \ldots, v_{12}$. Then $A$ is a $12 \times 12$ matrix with entries 0 or 1 and $a_{ij} = 1$ if $v_i$ and $v_j$ are joined by an edge, and

$a_{ij} = 0$ otherwise. Let $I$ be the $12 \times 12$ unit matrix and $J$ be the $12 \times 12$ all one matrix. Then the rows of the $12 \times 24$ matrix $(I \quad J-A)$ generate the extended binary Golay code.

**Exercise 2.4** Prove this. (First observe that the generated code is self-dual and that $(J - A^t \quad I)$ generates the same code. This allows one to study only linear combinations of 2 or 3 basis vectors. Consider first the sum of two rows of the generator matrix corresponding to two distinct vertices $v_j$ and $v_k$ of the icosahedron. There is a one at position $12+i$ $(i = 1, 2, \ldots, 12)$ of this vector if and only if the vertex $v_i$ of the icosahedron is joined by an edge to $v_j$ or $v_k$ but not to both of these vertices. Since there are exactly two 1's among the first 12 components of this vector, one has to show that for any choice of distinct vertices $v_j$ and $v_k$ there are at least 6 such vertices. To show this, use the symmetry of the icosahedron. Similarly, consider the sum of three rows corresponding to three distinct vertices $v_j$, $v_k$, and $v_l$. There is a one at position $12+i$ of this vector if and only if the vertex $v_i$ is either adjacent to exactly two or to none of the vertices $v_j$, $v_k$, and $v_l$. In this case one has to show that there are at least 5 such vertices.)

We have already seen that the automorphism group of the extended binary Golay code $\widetilde{G}$ is the automorphism group of the Steiner system $S(5,8,24)$, and this is the Mathieu group $M_{24}$.

**Proposition 2.10.** *The automorphism group of the extended binary Golay code $\widetilde{G}$ acts transitively on the* 24 *coordinate positions.*

*Proof.* Let $A$ be the adjacency matrix of the icosahedron. It follows from the above remarks that $(I \quad J-A)$ and $(J - A^t \quad I)$ both generate $\widetilde{G}$. Since the symmetry group of the icosahedron is transitive on the 12 vertices, the claim follows. □

**Corollary 2.3.** *Let $C$ be a binary code containing* 0 *with word length* 23*, minimum distance* 7*, and $|C| \geq 2^{12}$. Then $C$ is the binary Golay code.*

*Proof.* Extend $C$ by a parity check bit to obtain $\widetilde{G}$. Thus $C$ is obtained from $\widetilde{G}$ by suppressing some coordinate position, but all positions are equivalent, since the automorphism group acts transitively on the coordinate positions by Proposition 2.10. □

We have already seen that the Hamming weight enumerators of the codes $G$ and $\widetilde{G}$ are equal to

$$\begin{aligned} W_G(X,Y) &= X^{23} + 253X^{16}Y^7 + 506X^{15}Y^8 + 1288X^{12}Y^{11} \\ &\quad + 1288X^{11}Y^{12} + 506X^8Y^{15} + 253X^7Y^{16} + Y^{23}, \end{aligned}$$

$$W_{\widetilde{G}}(X,Y) = X^{24} + 759X^{16}Y^8 + 2576X^{12}Y^{12} + 759X^8Y^{16} + Y^{24}.$$

Now we want to construct an even unimodular lattice with $a_1 = 0$, i.e., which contains no roots, using the extended Golay code $\widetilde{G}$. Let us first consider the lattice $\Gamma_{\widetilde{G}} = \frac{1}{\sqrt{2}}\rho^{-1}(\widetilde{G})$, where $\rho : \mathbb{Z}^{24} \to \mathbb{F}_2^{24}$ is the reduction mod 2. Let us put for brevity $\Gamma = \rho^{-1}(\widetilde{G})$. Then

$$a_1 = 48$$

and the 48 roots correspond to the vectors of type $(\pm 2)^1 0^{23}$ of $\Gamma$. Here the notation $a^k b^l$ means $k$ components of type $a$ and $l$ components of type $b$. Proposition 2.8 yields

$$a_2 = 196560 - 24a_1 = 195408.$$

The vectors $x \in \Gamma$ with $x^2 = 8$ are the $759 \cdot 2^8$ vectors of type $(\pm 1)^8 0^{16}$ where $(\pm 1)^8$ is a contribution of $\widetilde{G}$, and the $\binom{24}{2} \cdot 4$ vectors of type $(\pm 2)^2 0^{22}$. They yield 195408 vectors of type $\Gamma_{\widetilde{G}}$ of squared length 4.

Now $x \in \Gamma = \rho^{-1}(\widetilde{G})$ can be written as $x = c + 2y$ with $c \in \widetilde{G}$ and $y \in \mathbb{Z}^{24}$. Since $\widetilde{G}$ is doubly even, $\sum x_i \in 2\mathbb{Z}$. Therefore we can define a homomorphism

$$\begin{aligned} \alpha : \Gamma &\longrightarrow \mathbb{F}_2 \\ x &\longmapsto \frac{1}{2} \sum x_i \pmod 2. \end{aligned}$$

Note that $\sum x_i \equiv 2 \sum y_i \pmod 4$ implies that

$$\frac{1}{2} \sum x_i \equiv \sum y_i \pmod 2.$$

Now $A := \alpha^{-1}(0)$ is a sublattice of $\Gamma$ of index 2 which contains no vectors of squared length 4. The lattice $\Gamma$ can be written as the disjoint union

$$\Gamma = A \cup N$$

of $A$ with the set $N := \alpha^{-1}(1)$. Then the set $A \cup \left(\frac{1}{2} \cdot \mathbf{1} + N\right)$, $\mathbf{1} := (1, \ldots, 1) \in \mathbb{Z}^{24}$, is also a lattice. Now an element of $\left(\frac{1}{2} \cdot \mathbf{1} + N\right)$ is of the form $\frac{1}{2} \cdot \mathbf{1} + c + 2y$, with $\sum y_i$ odd. Then

$$\left(\frac{1}{2} \cdot \mathbf{1} + c + 2y\right)^2 \equiv 6 + 2 \cdot (\mathbf{1} \cdot y) \equiv 6 + 2 \cdot \left(\sum y_i\right) \equiv 0 \pmod 4.$$

So we can make the following definition:

**Definition.** The *Leech lattice* is the lattice

$$\Lambda_{24} := \frac{1}{\sqrt{2}} \left( A \cup \left( \frac{1}{2} \cdot \mathbf{1} + N \right) \right).$$

**Lemma 2.7.** *The lattice $\Lambda_{24}$ is an even unimodular lattice, which contains no roots.*

*Proof.* That $\Lambda_{24}$ is even, follows from the above arguments.

Note that the lattice $A_1 := \frac{1}{\sqrt{2}} A$ is a sublattice of the unimodular lattice $\Gamma_{\widetilde{G}}$ of index 2. Since it has also index 2 in $\Lambda_{24}$, it follows that $\Lambda_{24}$ is unimodular.

Since $A$ contains no vectors of squared length 4 and the shortest vectors in $\left(\frac{1}{2} \cdot \mathbf{1} + \mathbb{Z}^{24}\right)$ are of the form $\left(\pm\frac{1}{2}, \ldots, \pm\frac{1}{2}\right)$ and therefore have length 6, we see that $\Lambda_{24}$ contains no roots. This proves the lemma. □

We finally indicate the 196560 vectors of squared length 4 in $\Lambda_{24}$. We shall indicate the corresponding vectors of squared length 8 in $A \cup \left(\frac{1}{2} \cdot \mathbf{1} + N\right)$:

| type | | number |
|---|---|---|
| $(\pm 1)^8 0^{16}$ | (in $A$) | $759 \cdot 2^7 = 97152$ |
| $(\pm 2)^2 0^{22}$ | (in $A$) | $4 \cdot \binom{24}{2} = 1104$ |
| $\left(\pm\frac{1}{2}\right)^{23}\left(\pm\frac{3}{2}\right)^{1}$ | ( in $\frac{1}{2}\mathbf{1}+N$) | $2^{12} \cdot 24 = 98304$ |
| | | total $= 196560$ |

## 2.9 The MacWilliams Identity and Gleason's Theorem

In this section we want to derive the MacWilliams identity for binary linear codes and Gleason's theorem from the theory of modular forms.

First let us consider the following theta function. Let

$$A(\tau) := \sum_{x \in \mathbb{Z}} q^{x \cdot x} = 1 + 2q + 2q^4 + 2q^9 + \dots .$$

This is the theta function of the lattice $\Gamma = \sqrt{2}\mathbb{Z}$. So

$$A(\tau) = \sum_{x \in \Gamma} q^{\frac{1}{2}(x \cdot x)} = \sum_{x \in 2\mathbb{Z}} q^{\frac{1}{4}(x \cdot x)},$$

Consider also

$$B(\tau) := \sum_{x \in 2\mathbb{Z}+1} q^{\frac{1}{4}(x \cdot x)}.$$

This is not a theta function of a lattice. Then

$$A(\tau) + B(\tau) = \sum_{x \in \mathbb{Z}} q^{\frac{1}{4}(x \cdot x)} = \sum_{x \in \frac{1}{\sqrt{2}}\mathbb{Z}} q^{\frac{1}{2}(x \cdot x)}.$$

But the lattice $\frac{1}{\sqrt{2}}\mathbb{Z}$ is the dual lattice of $\Gamma = \sqrt{2}\mathbb{Z}$. Hence from Proposition 2.1 we derive the following transformation formula

$$A\left(-\frac{1}{\tau}\right) = \left(\frac{\tau}{i}\right)^{1/2} \frac{1}{\sqrt{2}} (A(\tau) + B(\tau)).$$

Replacing $\tau$ by $-\frac{1}{\tau}$, we get from this formula

$$B\left(-\frac{1}{\tau}\right) = \left(\frac{\tau}{i}\right)^{1/2} \frac{1}{\sqrt{2}} (A(\tau) - B(\tau)).$$

The matrix

$$\frac{1}{\sqrt{2}} \begin{pmatrix} 1 & 1 \\ 1 & -1 \end{pmatrix}$$

defines a transformation in the $(A,B)$-plane which is a rotation by $45^\circ$ followed by a reflection. The following homogeneous polynomial in $A$ and $B$ of degree 24 is invariant under this transformation:

$$A^4B^4\left(A^2-B^2\right)^4\left(A^2+B^2\right)^4 = A^4B^4\left(A^4-B^4\right)^4.$$

Since $A^4B^4\left(A^4-B^4\right)^4$ is also invariant under the transformation $\tau \longmapsto \tau+1$, it is a modular form of weight 12. Now

$$B(\tau) = \sum_{\substack{x\in 2\mathbb{Z}+1\\ x=2y+1}} q^{\frac{1}{4}(4y^2+4y+1)} = q^{\frac{1}{4}}\sum_{y\in\mathbb{Z}} q^{y^2+y} = 2q^{\frac{1}{4}} + \text{ higher order terms.}$$

Thus

$$B^4(\tau) = 16q + \text{ higher order terms,}$$

$$A^4B^4\left(A^4-B^4\right)^4 = 16q + \text{ higher order terms,}$$

and hence

$$A^4B^4\left(A^4-B^4\right)^4 = 16\Delta = 16q\prod_{n=1}^{\infty}(1-q^n)^{24}.$$

**Proposition 2.11.** *Let $C \subset \mathbb{F}_2^n$ be a binary linear code with Hamming weight enumerator $W_C(X,Y)$. Then*

$$\vartheta_{\Gamma_C} = W_C(A,B).$$

*Proof.* Consider a codeword $c \in C$, and let $\rho : \mathbb{Z}^n \to \mathbb{F}_2^n$ be the reduction mod 2. Then

$$\sum_{x\in\frac{1}{\sqrt{2}}\rho^{-1}(c)} q^{\frac{1}{2}(x\cdot x)} = \sum_{x\in\rho^{-1}(c)} q^{\frac{1}{4}(x\cdot x)} = A^{n-w(c)}B^{w(c)},$$

by the definition of $A$ and $B$. This proves the proposition. □

**Example 2.3** For the extended Hamming code $\widetilde{H}$ we have

$$W_{\widetilde{H}}(X,Y) = X^8 + 14X^4Y^4 + Y^8$$

and hence

$$\vartheta_{E_8} = E_4 = A^8 + 14A^4B^4 + B^8.$$

**Theorem 2.7 (MacWilliams).** *Let $C \subset \mathbb{F}_2^n$ be a binary linear code of type $[n,k,d]$. Then*

$$W_{C^\perp}(X,Y) = \frac{1}{2^k}W_C(X+Y,X-Y).$$

*Proof.* We derive this identity from the corresponding identity for theta functions (Proposition 2.1) using Proposition 2.11. We have

$$W_C\left(A\left(-\frac{1}{\tau}\right),B\left(-\frac{1}{\tau}\right)\right)$$

$$
\begin{aligned}
&= \vartheta_{\Gamma_C}\left(-\frac{1}{\tau}\right) \\
&= \left(\frac{\tau}{i}\right)^{\frac{n}{2}} \frac{1}{2^{\frac{n}{2}-k}} \vartheta_{\Gamma_C^*}(\tau) \qquad \text{(by Proposition 2.1)} \\
&= \left(\frac{\tau}{i}\right)^{\frac{n}{2}} \frac{1}{2^{\frac{n}{2}-k}} \vartheta_{\Gamma_{C^\perp}}(\tau) \qquad \text{(by Lemma 1.11)} \\
&= \left(\frac{\tau}{i}\right)^{\frac{n}{2}} \frac{1}{2^{\frac{n}{2}-k}} W_{C^\perp}(A(\tau),B(\tau)).
\end{aligned}
$$

By the transformation formulas of $A$ and $B$, we have on the other hand

$$
W_C\left(A\left(-\frac{1}{\tau}\right),B\left(-\frac{1}{\tau}\right)\right) = \left(\frac{\tau}{i}\right)^{n/2} \frac{1}{2^{n/2}} W_C(A(\tau)+B(\tau),A(\tau)-B(\tau)).
$$

Thus

$$
W_{C^\perp}(A,B) = \frac{1}{2^k} W_C(A+B,A-B).
$$

We know from Corollary 2.2 that the algebra of modular forms is isomorphic to the polynomial algebra $\mathbb{C}[E_4,E_6]$. This implies that the modular forms of weight divisible by 4 form a polynomial algebra $\mathbb{C}[E_4,\Delta]$ generated by $E_4$ and $\Delta$. But

$$
E_4 = A^8 + 14A^4B^4 + B^8,
$$

$$
\Delta = \frac{1}{16} A^4B^4(A^4-B^4)^4.
$$

Therefore $A$ and $B$ are algebraically independent over $\mathbb{C}$. This proves Theorem 2.7. □

**Corollary 2.4.** *If $C \subset \mathbb{F}_2^n$ is a self-dual code, then*

$$
W_C(X,Y) = W_C\left(\frac{X+Y}{\sqrt{2}}, \frac{X-Y}{\sqrt{2}}\right).
$$

*This means that the Hamming weight enumerator is invariant under a rotation by 45° followed by a reflection.*

**Example 2.4** For $n=2$, $k=1$, $C=\{(0,0),(1,1)\} \subset \mathbb{F}_2^2$ we have

$$
W_C(X,Y) = X^2+Y^2.
$$

This polynomial is invariant under rotations.

**Theorem 2.8 (Gleason).** *Let $C \subset \mathbb{F}_2^n$ be a doubly even self-dual code. Then the Hamming weight enumerator $W_C(X,Y)$ is a polynomial in*

$$
\varphi := W_{\widetilde{H}}(X,Y) = X^8 + 14X^4Y^4 + Y^8
$$

*and*

$$
\xi := X^4Y^4\left(X^4-Y^4\right)^4
$$

*or, equivalently, in the weight enumerator $W_{\widetilde{H}}$ of the extended Hamming code and the weight enumerator*

$$W_{\widetilde{G}}(X,Y) = \left(X^8 + 14X^4Y^4 + Y^8\right)^3 - 42X^4Y^4\left(X^4 - Y^4\right)^4$$

*of the extended Golay code.*

*Proof.* $W_C(A,B)$ is a modular form of weight $\frac{n}{2}$. Since $C$ is doubly even and self-dual, $\frac{n}{2}$ is divisible by 4. As we have already seen in the proof of Theorem 2.7, every modular form of weight divisible by 4 is a polynomial in $E_4$ and $\Delta$. Hence $W_C(A,B)$ is a polynomial in $A^8 + 14A^4B^4 + B^8$ and $A^4B^4\left(A^4 - B^4\right)^4$. This proves Theorem 2.8. □

Theorems 2.7 and 2.8 can also be proved (and were first proved) without using the theory of modular forms.

Analogues of Theorems 2.7 and 2.8 for so-called local weight enumerators were obtained in [74] using theta functions with spherical coefficients (which are considered in Sect. 3.1).

**Example 2.5** Let $n = 32$, and $C \subset \mathbb{F}_2^{32}$ a doubly even self-dual code. Then, by Theorem 2.8,

$$\begin{aligned} W_C(X,Y) = &\left(X^8 + 14X^4Y^4 + Y^8\right)^4 \\ &+ bX^4Y^4\left(X^4 - Y^4\right)^4\left(X^8 + 14X^4Y^4 + Y^8\right). \end{aligned}$$

If this code has minimum distance 8, i.e., $A_4 = 0$, where $A_4$ is the coefficient of $X^{28}Y^4$, then $b = -56$. Thus

$$A_8 = 14^2 \cdot 6 + 4 - 560 = 620.$$

Now let $n = 24m + 8k$, $k = 0,1,2$. Let $C \subset \mathbb{F}_2^n$ be a doubly even self-dual code with weight enumerator $W_C(X,Y)$. By Theorem 2.8, $W_C$ can be written as

$$W_C = \sum_{j=0}^{m} b_j \varphi^{3(m-j)+k} \xi^j, \qquad b_j \in \mathbb{C}.$$

Now suppose that the $b_j$ are chosen so that $W_C(X,Y)$ has as many leading coefficients equal to zero as possible. Then also the first possibly nonzero coefficient, the coefficient of $X^{n-4m-4}Y^{4m+4}$ which we denote by $A^*_{4m+4}$, is uniquely determined. So

$$W_C(X,Y) = X^n + A^*_{4m+4}X^{n-4m-4}Y^{4m+4} + \dots.$$

The resulting $W_C$ is the weight enumerator of a doubly even self-dual code with the greatest minimum weight we might hope to attain, and is called an *extremal* weight enumerator. A linear code having this weight enumerator is called an *extremal code*. An extremal code has minimum distance $d \geq 4m + 4$. In fact it has minimum distance $d = 4m + 4$, since it can be shown that $A^*_{4m+4} \neq 0$ for all $m \geq 1$. For example, for $k = 0$ one has the following formula for $A^*_{4m+4}$.

**Theorem 2.9.** *For an extremal doubly even self-dual code $C \subset \mathbb{F}_2^n$, $n = 24m$, one has*

$$A^*_{4m+4} = \binom{n}{5}\binom{5m-2}{m-1} \Big/ \binom{4m+4}{5}.$$

*Proof.* This follows from the fact that the codewords of $C$ of weight $4m+4$ form a 5-design (Theorem of Assmus and Mattson, cf. [66, Chap. 6, §4, Theorem 9]). For details see [66, Chap. 19, §5]. □

**Example 2.6** For $m = 1 : A^*_8 = 759$, for $m = 2 : A^*_{12} = 17296$.

But, by a result of Mallows, one has

$$A^*_{4m+8} < 0$$

for $n = 24m$ sufficiently large. This shows that extremal doubly even self-dual codes of length $n = 24m$ can only exist for finitely many $m$. A computer was used to show that $A^*_{4m+8}$ first becomes negative at around $n = 3720$:

$$A^*_{624} \approx 1.163 \cdot 10^{170},$$
$$A^*_{628} \approx -5.848 \cdot 10^{170}.$$

In fact one knows the existence of extremal doubly even self-dual codes for $n \leq 64$ and some $n > 72$. It is not known whether there exists such a code for $n = 72$. For more details see [66, Chap. 19, §5].

There are similar results for lattices. Let $\Gamma$ be an even unimodular lattice in $\mathbb{R}^n$, $n = 24m + 8k$, $k = 0, 1$, or $2$. The theta function $\vartheta_\Gamma$ can be written as

$$\vartheta_\Gamma = \sum_{j=0}^{m} b_j E_4^{3(m-j)+k} \Delta^j, \qquad b_j \in \mathbb{C}.$$

If

$$\vartheta_\Gamma(\tau) = 1 + a^*_{2m+2} q^{2m+2} + \dots,$$

then this theta function is called an *extremal* theta function and a corresponding lattice is called an *extremal* lattice. For an extremal even unimodular lattice, one can also show that $a^*_{2m+2} > 0$ for all $m$, but $a^*_{2m+4} < 0$ for all sufficiently large $m$. So like extremal codes, extremal even unimodular lattices can only exist for finitely many $n$. The coefficient $a^*_{2m+4}$ first becomes negative at around $n = 41000$. The existence of an extremal even unimodular lattice in $\mathbb{R}^{72}$ was only proven in 2010 by G. Nebe [67]. For more details see [21, Chap. 7, §7].

## 2.10 Quadratic Residue Codes

By Example 2.5 an extremal doubly even self-dual code of length 32 would have minimum distance $d = 8$ and $A_8 = 620$. Similarly, an extremal doubly even self-dual code of length 48 would have minimum distance $d = 12$ and $A_{12} = 17296$ (cf. Example 2.6). So far, we do not know the existence of such codes. Our aim in this section is to construct such codes. This construction will be part of a general construction of an important class of codes, the so called quadratic residue codes. A general reference for this section is [56, Chap. 6].

Let $p$ be a prime number. Then we denote the field $\mathbb{Z}/p\mathbb{Z}$ by $\mathbb{F}_p$. If $\mathbb{F}_q$ is any finite field with $q$ elements then $q$ must be a power of some prime number $p$, so $q = p^r$ for some $r$. The number $q$ determines $\mathbb{F}_q$ up to isomorphism: The multiplicative group $\mathbb{F}_q^*$ is a cyclic group of order $q-1$, and the elements of $\mathbb{F}_q$ are exactly the $q$ distinct zeros of the polynomial $x^q - x$. Therefore $\mathbb{F}_q$ is the splitting field of $x^q - x$ and therefore uniquely determined by $q$.

The mapping $\sigma : \mathbb{F}_q \to \mathbb{F}_q$, $x \mapsto x^p$, is a field automorphism of $\mathbb{F}_q$ which fixes the subfield $\mathbb{F}_p$. We have $\sigma^r = 1$. The automorphism $\sigma$ generates the Galois group of $\mathbb{F}_q$ over $\mathbb{F}_p$ which is therefore cyclic of order $r$. If $s$ divides $r$ then $\sigma^s$ generates a cyclic subgroup of order $r/s$. The subfield $\mathrm{Fix}(\sigma^s)$ of $\mathbb{F}_q$ fixed by $\sigma^s$ consists of the zeros of the polynomial $x^{p^s} - x$. Hence $\mathrm{Fix}(\sigma^s) = \mathbb{F}_{p^s} \subset \mathbb{F}_{p^r}$. The polynomial $x^{p^r} - x$ is the product of all irreducible polynomials over $\mathbb{F}_p$ of degree $s$ with $s|r$ which are *monic*, i.e., where the coefficient of $x^s$ is 1. The orbits of the Galois group of $\mathbb{F}_{p^r}$ over $\mathbb{F}_p$ are just the zeros of the irreducible factors of $x^{p^r} - x$.

**Example 2.7** Let $p = 2$, $r = 4$. Then $|\mathbb{F}_{2^4}^*| = 15$ and

$$x^{16} - x = x(x+1)(x^2+x+1)(x^4+x+1)(x^4+x^3+1)(x^4+x^3+x^2+x+1).$$

Let $\alpha$ be a generator of $\mathbb{F}_{2^4}^*$. The orbits of the Galois group of $\mathbb{F}_{2^4}$ over $\mathbb{F}_2$ are

$$\{1\},\ \{\alpha,\alpha^2,\alpha^4,\alpha^8\},\ \{\alpha^3,\alpha^6,\alpha^{12},\alpha^9\}\ \{\alpha^5,\alpha^{10}\},\ \{\alpha^7,\alpha^{14},\alpha^{13},\alpha^{11}\}.$$

Let $k = \mathbb{F}_q$ be a finite field with $q = p^r$ elements. We denote by $k[x]$ the ring of polynomials over $k$. Let $n$ be a natural number with $(n,p) = 1$ and consider the polynomial $x^n - 1$ over $k$. The multiples of $x^n - 1$ form a principal ideal in $k[x]$ which we denote by $(x^n - 1)$. The elements of the quotient ring $k[x]/(x^n - 1)$ are represented by polynomials $a_0 + a_1x + \ldots + a_{n-1}x^{n-1}$ with $a_i \in k$ for $0 \le i \le n-1$. We therefore identify $k[x]/(x^n - 1)$ with $k^n$ by mapping $a_0 + a_1x + \ldots + a_{n-1}x^{n-1}$ to its $n$-tuple of coefficients $(a_0, \ldots, a_{n-1})$.

Now let $x^n - 1 = g(x)h(x)$, where $g(x), h(x) \in k[x]$ and $\deg(g) = m < n$ and consider the code

$$C = \{a(x)g(x) (\mathrm{mod}\ x^n - 1) \,|\, a(x) \in k[x]\} \subset k^n.$$

Such a code is called a *cyclic code*. This code is a linear code of length $n$ and dimension $n - m$. The reason for the name cyclic code is the following: If $C$ is a cyclic code and $c(x) = c_0 + c_1x + \ldots + c_{n-1}x^{n-1}$ is any codeword, then $xc(x)$ is also a codeword, i.e.,

$$(c_{n-1}, c_0, c_1, \ldots, c_{n-2}) \in C\,.$$

Conversely, one easily sees that a linear code $C$ having this property corresponds to an ideal in $k[x]/(x^n - 1)$ and therefore is cyclic.

The polynomial $g(x)$ is called the *generator polynomial* of $C$. Write $g(x) = b_0 + b_1 x + \ldots + b_{n-1}x^{n-1}$ with $b_{m+1} = \ldots = b_{n-1} = 0$. The codewords $g(x)$, $xg(x)$, $\ldots$ , $x^{n-1}g(x)$ generate $C$, hence a generator matrix of $C$ is given by the first $n-m$ rows of the matrix

$$\begin{pmatrix} b_0 & b_1 & \ldots & b_{n-1} \\ b_{n-1} & b_0 & \ldots & b_{n-2} \\ \vdots & \vdots & \ddots & \vdots \\ b_1 & b_2 & \ldots & b_0 \end{pmatrix}.$$

It is in general not so easy to determine the minimum distance $d$ of a cyclic code. We now assume that

$$g(x) = (x-\alpha)(x-\alpha^2)\cdots(x-\alpha^r)\widetilde{g}(x)$$

in $k[x]$, where $\alpha$ is a primitive $n$-th root of unity. A cyclic code with such a generator polynomial is called a *BCH code*, because it was first considered by R. C. Bose, D. K. Ray-Chaudhuri, and A. Hocquenghem.

**Proposition 2.12.** *The minimum distance $d$ of $C$ is at least $r+1$.*

*Proof.* Assume that $a_{t_1}x^{t_1} + \ldots + a_{t_r}x^{t_r}$ is a codeword in $C$, where the $t_i$ are pairwise distinct integers with $0 \le t_i \le n-1$ for $i = 1, \ldots, r$. Then we must have

$$\begin{aligned} a_{t_1}\alpha^{t_1} + \ldots + a_{t_r}\alpha^{t_r} &= 0 \\ a_{t_1}\alpha^{2t_1} + \ldots + a_{t_r}\alpha^{2t_r} &= 0 \\ &\vdots\;\vdots\;\vdots \\ a_{t_1}\alpha^{rt_1} + \ldots + a_{t_r}\alpha^{rt_r} &= 0. \end{aligned}$$

These are $r$ linear equations for $a_{t_1}, \ldots, a_{t_r}$. The determinant of the coefficient matrix is

$$\begin{aligned} \begin{vmatrix} \alpha^{t_1} & \ldots & \alpha^{t_r} \\ \alpha^{2t_1} & \ldots & \alpha^{2t_r} \\ \vdots & \ddots & \vdots \\ \alpha^{rt_1} & \ldots & \alpha^{rt_r} \end{vmatrix} &= \alpha^{t_1}\cdots\alpha^{t_r} \begin{vmatrix} 1 & \ldots & 1 \\ \alpha^{t_1} & \ldots & \alpha^{t_r} \\ \vdots & \ddots & \vdots \\ (\alpha^{t_1})^{r-1} & \ldots & (\alpha^{t_r})^{r-1} \end{vmatrix} \\ &= \pm\alpha^{t_1}\cdots\alpha^{t_r} \cdot \prod_{i<j}(\alpha^{t_i} - \alpha^{t_j}), \end{aligned}$$

since the last determinant is a Vandermonde determinant. Since $\alpha$ is a primitive $n$-th root of unity, this determinant is different from zero. Therefore the above system of linear equations has only the trivial solution and Proposition 2.12 follows. □

**Remark 2.3** The bound of Proposition 2.12 is usually called the *BCH bound*.

Note that if $k_0 \subset k$ is a subfield of $k$ and $g(x)$ lies in $k_0[x]$ then we can consider everything over $k_0$. The corresponding linear code $C \subset k_0^n$ also has dimension $n - \deg(g)$ over $k_0$.

Now let $p = 2$, $q = 2^m$, and $n = q - 1 = 2^m - 1$. Furthermore, let $\alpha$ be a primitive $n$-th root of unity, and let $g(x) \in \mathbb{F}_2[x]$ be an irreducible polynomial with $g(\alpha) = 0$. By the remarks at the beginning of this section, $g(x)$ has degree $m$ and

$$g(x) = (x - \alpha)(x - \alpha^2)(x - \alpha^4) \cdots (x - \alpha^{2^{m-1}}).$$

The corresponding linear code $C_{(q)} \subset \mathbb{F}_q^n$ is called a (generalized) *Hamming code*. By Proposition 2.12 it has minimum distance at least 3. Since $g(x) \in \mathbb{F}_2[x]$ we can also consider the linear code $C \subset \mathbb{F}_2^n$ over $\mathbb{F}_2$ of the same length $n$ and dimension $n - m$.

**Proposition 2.13.** *The linear code $C$ is a perfect code with minimum distance* 3.

*Proof.* If $C$ is an $[n,k,d]$-code with $d \geq 2t+1$ then spheres of radius $t$ around codewords are pairwise disjoint. This implies that

$$2^k \left[1 + \binom{n}{1} + \ldots + \binom{n}{t}\right] \leq 2^n.$$

In our case we have

$$2^{n-m}[1+n] = 2^n.$$

Therefore $d$ must be less than or equal to 4. Assume now that $d = 4$. Let $c \in C$ be a codeword of weight 4. Then there exists $x \notin C$ of weight 2 with $d(x,c) = 2$. Since the spheres of radius 1 around codewords cover $\mathbb{F}_2^n$, there exists a $c' \in C$ with $d(x,c') = 1$. But then $d(c,c') \leq 3$, a contradiction. Therefore the minimum distance of $C$ is 3 and by the definition of Sect. 2.8 $C$ is a perfect code. This proves Proposition 2.13. □

For $m = 3$, $C$ is a binary $[7,4,3]$-code and hence equal to the $[7,4,3]$-Hamming code $H$ introduced in Sect. 1.2. For $m = 2$ we get the repetition code of length 3 considered in the same section.

We want to determine the dual code to a Hamming code. For that purpose let $x^n - 1 = g(x)h(x)$ in $\mathbb{F}_2[x]$, where $n$ is odd. Denote by $C_g$ the cyclic code with generator polynomial $g(x)$.

**Lemma 2.8.** *The dual code $C_g^\perp$ is equal to the cyclic code $C_{\widetilde{h}}$ generated by the polynomial $\widetilde{h}(x) \equiv x^{\deg h} h(x^{-1}) \pmod{x^n - 1}$.*

*Proof.* Since $\deg \widetilde{h} = \deg h$ and $\deg g + \deg h = n$, we have $\dim C_g + \dim C_{\widetilde{h}} = n$. In order to show that $C_{\widetilde{h}} \subset C_g^\perp$ note that the inner product $a \cdot b$ of two elements $a(x), b(x) \in \mathbb{F}_2[x]/(x^n - 1)$ is the constant term of $a(x)b(x^{-1})$. But $g(x)\widetilde{h}(x^{-1}) = g(x)h(x)x^{-\deg h} = (x^n - 1)x^{m-n}$ has constant term 0. This proves Lemma 2.8. □

A linear code $C \subset \mathbb{F}_2^n$ is called a *simplex code* if and only if all nonzero codewords have the same weight $d$. For the weight enumerator of a simplex code $C$ of dimension $m$ we have

$$W_c(X,Y) = X^n + aX^{n-d}Y^d,$$

where $a = 2^m - 1$.

**Lemma 2.9.** *Let $C$ be a simplex code, and let $n = 2^m - 1$. Then $C^\perp$ has minimum distance $\geq 2$ if and only if $d = \frac{n+1}{2}(= 2^{m-1})$.*

*Proof.* By the MacWilliams identity (Theorem 2.7) we have

$$W_{C^\perp}(X,Y) = \frac{1}{2^m}((X+Y)^n + a(X+Y)^{n-d}(X-Y)^d).$$

The coefficient of $X^{n-1}Y$ is

$$\frac{1}{2^m}(n + n(n-d-d)) = \frac{1}{2^m}n(n+1-2d).$$

Lemma 2.9 follows from this. □

Now consider a Hamming code $C$. This is an $[n, n-m, 3]$-code over $\mathbb{F}_2$ with $n = 2^m - 1$.

**Exercise 2.5** Show that the minimum distance of the dual code $C^\perp$ is $\geq 2^{m-1}$ by using Proposition 2.12.

**Proposition 2.14.** *The dual code $C^\perp$ of a Hamming code is a simplex code with minimum distance $2^{m-1}$.*

*Proof.* The code $C^\perp$ is a linear code of length $n$ and dimension $m$. By Lemma 2.8 the generator polynomial of $C^\perp$ is $\widetilde{h}$, $\deg(\widetilde{h}) = n - m$. Then the codewords

$$0, \widetilde{h}(x), x\widetilde{h}(x), \ldots, x^{n-1}\widetilde{h}(x)$$

are contained in $C^\perp$. We claim that they are all different in $\mathbb{F}_2[x]/(x^n - 1)$. Otherwise, a difference $(x^r - x^s)\widetilde{h}(x)$ must be divisible by $x^n - 1$, hence must vanish at all roots of unity. In particular $\alpha^{-1}$ must be a zero of $x^r - x^s$, i.e., $\alpha^{-r} - \alpha^{-s} = 0$ for $r, s < n$, $r \neq s$, which is absurd. Therefore $C^\perp$ has exactly $2^m$ codewords, all having the same weight. Hence $C^\perp$ is a simplex code, and by Lemma 2.9 its minimum distance is equal to $2^{m-1}$. This proves Proposition 2.14. □

In particular the dual $H^\perp$ of the $[7,4,3]$-Hamming code $H$ is a simplex code; this code was considered in Sect. 1.4.

We consider again a general cyclic code over a finite field $k$ of characteristic $p > 0$. For a natural number $n$ with $(n, p) = 1$ let

$$x^n - 1 = g(x)h(x),$$

and let $C$ be the cyclic code in $k[x]/(x^n - 1)$ generated by $g(x)$. Since $x^n - 1$ has no multiple zeros, the polynomials $g(x)$ and $h(x)$ have no common divisor in $k[x]$. Hence there exist polynomials $a(x), b(x) \in k[x]$ such that $a(x)g(x) + b(x)h(x) = 1$. Define

$$c(x) := a(x)g(x) \pmod{x^n - 1}.$$

Then $c(x)$ is a codeword in $C$ which satisfies $c^2(x) = c(x)$. Therefore $c(x)$ is called an *idempotent* of $C$. Moreover, $c(x)g(x) = g(x)$ implies that $c(x)p(x) = p(x)$ for all $p(x) \in C$.

So $c(x)$ is an identity element for $C$. Now let $\alpha$ be an $n$-th root of unity in an extension field of $k$. Then $c(\alpha) = c(\alpha)c(\alpha)$ and hence $c(\alpha) = 0$ or $c(\alpha) = 1$. It follows that $c(\alpha) = 0$ if $g(\alpha) = 0$ and $c(\alpha) = 1$ if $h(\alpha) = 0$ for all $n$-th roots of unity $\alpha$. There is a unique codeword $c(x) \in C$ with $c^2(x) = c(x)$ which satisfies this condition: For if $c(x)$ is such a codeword, then $c(x)$ has to be a multiple of $g(x)$ and $1-c(x)$ a multiple of $h(x)$ in $k[x]$. But this means that there exist polynomials $\widetilde{a}(x)$ and $\widetilde{b}(x)$ such that $1 = \widetilde{a}(x)g(x) + \widetilde{b}(x)h(x)$. This implies that $a(x) - \widetilde{a}(x)$ is a multiple of $h(x)$, hence $c(x)$ is uniquely determined modulo $x^n - 1$.

Since $c(x)g(x) = g(x)$, we see that $c(x)$ generates the ideal $C$. Hence the codewords obtained by cyclic permutations of the coefficients of $c(x)$ generate the code $C$.

We shall now introduce quadratic residue codes. For that purpose we assume that the prime $p$ satisfies $p \equiv 7 \pmod 8$. Then $\mathbb{F}_p = \mathbb{Z}/p\mathbb{Z} = \{0\} \cup Q \cup N$ where $Q$ is the set of quadratic residues and $N$ is the set of non-squares in $\mathbb{F}_p$. A *quadratic residue* in $\mathbb{F}_p$ is the residue class of a square in $\mathbb{Z}$. The sets $Q$ and $N$ have the same number $\frac{p-1}{2}$ of elements. Now let $n := p$ and consider the polynomial $x^p - 1$ over $k = \mathbb{F}_2$. If $2^m - 1 \equiv 0 \pmod p$ (e.g. if $m = p-1$) then $x^p - 1$ splits over $\mathbb{F}_{2^m}$ into linear factors.

**Example 2.8** For $p = 23$ we can take $m = 11$, since $2^{11} - 1 = 2047 = 23 \cdot 89$. This implies that 2 is not a generator of $(\mathbb{F}_{23})^*$. But 5 is a generator. The powers of 5 in $\mathbb{F}_{23}$ are

$$5, 2, 10, 4, 20, 8, 17, 16, 11, 9, 22, 18, 21, 13, 19, 3, 15, 6, 7, 12, 14, 1.$$

Let $\alpha$ be a primitive $p$-th root of unity in $\mathbb{F}_{2^m}$. Then

$$x^p - 1 = \prod_{j=0}^{p-1} (x - \alpha^j).$$

Define

$$g(x) := \prod_{r \in Q} (x - \alpha^r),$$
$$h(x) := (x-1) \prod_{r \in N} (x - \alpha^r).$$

Then $x^p - 1 = g(x)h(x)$. We claim that $g(x)$ and $h(x)$ both have coefficients in $\mathbb{F}_2$. To prove this , we have to show that $a^2 = a$ for all coefficients, hence that

$$\prod_{r \in Q} (x - \alpha^{2r}) = \prod_{r \in Q} (x - \alpha^r)$$

and that the same identity holds with $Q$ replaced by $N$. But by assumption $p \equiv -1 \pmod 8$, and hence 2 is a quadratic residue. Therefore the claim follows.

**Definition.** The cyclic code $C$ of length $p$ over $\mathbb{F}_2$ with generator polynomial $g(x)$ is called a *(binary) quadratic residue code*.

We shall determine an idempotent for $C$. Define

$$c(x) := \sum_{r \in Q} x^r \in \mathbb{F}_2[x].$$

Since $p \equiv -1 \pmod 4$, we have $-1 \in N$. Therefore the substitution $\alpha \mapsto \alpha^{-1}$ changes the code, and the idempotent of $C$ must depend on $\alpha$. Clearly $c^2(x) = c(x)$ in $\mathbb{F}_2[x]/(x^p - 1)$. Therefore $c(\beta) = 0$ or $c(\beta) = 1$ for every $p$-th root of unity $\beta$. We have $c(1) = \frac{p-1}{2} = 1$. If $\beta$ denotes a $p$-th root of unity different from 1, then

$$c(\beta) + c(\beta^{-1}) = 1.$$

For if $\beta = \alpha^s$, $s \in Q$, then $\beta^{-1} = \alpha^{-s}$, $-s \in N$ and

$$\sum_{r \in Q \cup N} \beta^r = 1.$$

Moreover we have $c(\alpha^s) = c(\alpha)$ for $s \in Q$. Therefore we see that $c(x)$ is an idempotent for $C$ where we have to replace $\alpha$ by $\alpha^{-1}$ if necessary, yielding an isomorphic code.

The minimum distance of a quadratic residue code is in general not known.

**Proposition 2.15.** *Let $a \in C$ be a codeword with weight $w(a) = d$.*

(i) *If $d$ is odd (i.e., $a(1) \neq 0$), then $d^2 - d + 1 \geq p$ and $d \equiv 3 \pmod 4$.*

(ii) *If $d$ is even, then $d \equiv 0 \pmod 4$.*

*Proof.* Let $a(x) = \sum_{i=1}^{d} x^{k_i}$ where the $k_i$ are distinct elements of $\mathbb{Z}/p\mathbb{Z}$. Define $\widehat{a}(x) := a(x^{-1})$. Then $a(\alpha^r) = 0$ for $r \in Q$ and $\widehat{a}(\alpha^r) = 0$ for $r \in N$. First assume that $d$ is odd, so $a(1) \neq 0$. Then $a(x)\widehat{a}(x)$ is a multiple of the polynomial $1 + x + \ldots + x^{p-1}$ and hence equal to this polynomial in $\mathbb{F}_2[x]/(x^p - 1)$. It follows that the weight of $a \cdot \widehat{a}$ is $p$. On the other hand, by evaluating the product $a(x)\widehat{a}(x)$ we see that there are $d^2$ monomials and $d$ of them (of the form $x^{k_i}x^{-k_i}$) coincide. This implies that $d^2 - d + 1 \geq p$. Another type of cancellation may arise from a pair with the same exponents. But then there is always a corresponding pair with the same exponents but opposite signs. Hence, if terms in the product $a(x)\widehat{a}(x)$ cancel then they cancel four at a time. Therefore $d^2 - d + 1 - 4e = p$ for some $e \geq 0$. Since $d$ is odd, it follows that $d \equiv 3 \pmod 4$. This proves (i).

If $d$ is even then $a(x)\widehat{a}(x) = 0$ in $\mathbb{F}_2[x]/(x^p - 1)$. As above we obtain $d^2 - d - 4e = 0$ and hence $d \equiv 0 \pmod 4$. This proves (ii) and completes the proof of Proposition 2.15. □

Let $\widetilde{C} \subset \mathbb{F}_2^{p+1}$ be the corresponding extended code of $C$ (cf. Sect. 1.2). We have $\dim \widetilde{C} = \dim C = p - \frac{p-1}{2} = \frac{p+1}{2}$. By Proposition 2.15 $\widetilde{C}$ is doubly even and therefore $\widetilde{C} \subset \widetilde{C}^\perp$. It follows that $\widetilde{C}$ is also self-dual.

Let $c(x) = c_0 + c_1 x + \ldots + c_{p-1} x^{p-1}$ be the above idempotent of $C$. So $c_r = 1$ if $r \in Q$ and $c_r = 0$ if $r \notin Q$. Then the first $\frac{p+1}{2}$ rows of the matrix

$$G = \begin{pmatrix} 1 & 1 & 1 & \cdots & 1 \\ 1 & c_0 & c_1 & \cdots & c_{p-1} \\ 1 & c_{p-1} & c_0 & \cdots & c_{p-2} \\ \vdots & \vdots & \vdots & \ddots & \vdots \\ 1 & c_1 & c_2 & \cdots & c_0 \end{pmatrix}$$

form a generator matrix for $\widetilde{C} \subset \mathbb{F}_2^{p+1}$, where we have taken the overall parity check in front. We now number the coordinates of $\mathbb{F}_2^{p+1}$ by the points of the projective line $\mathbb{P}_1(\mathbb{F}_p) = \mathbb{F}_p \cup \{\infty\}$, i.e., by $\infty, 0, 1, \ldots, p-1$. The group $\mathrm{PGL}_2(\mathbb{F}_p)$ of fractional linear transformations $z \mapsto \frac{az+b}{cz+d}$, $a,b,c,d \in \mathbb{F}_p$, with $ad - bc \neq 0$ operates on $\mathbb{F}_2^{p+1}$ by permutation of coordinates. We consider the subgroup $\mathrm{PSL}_2(\mathbb{F}_p)$ of fractional linear transformations with $ad - bc = 1$.

**Proposition 2.16.** $\mathrm{PSL}_2(\mathbb{F}_p) \subset \mathrm{Aut}\,\widetilde{C}$.

*Proof.* (Cf. [55, Proof of Theorem (4.4.8)].) The group $\mathrm{PSL}_2(\mathbb{F}_p)$ is generated by the fractional linear transformations $S: z \mapsto z+1$ and $T: z \mapsto -\frac{1}{z}$. Clearly $S$ is a cyclic shift of $0, 1, \ldots, p-1$ and fixes $\infty$. So it leaves $\widetilde{C}$ invariant.

To study the action of $T$, number the rows of $G$ by $\infty, 0, 1, \ldots, p-1$. So $G = ((g_{ij}))_{i,j \in \{\infty, 0, 1, \ldots, p-1\}}$. Define $h_{ij} := g_{-\frac{1}{i}, -\frac{1}{j}}$, $H := ((h_{ij}))$. We have to show that the rows of $H$ belong to the code $\widetilde{C}$. Let $i \in \{\infty, 0, 1, \ldots, p-1\}$.

If $i = \infty$ then $h_{\infty,j} = g_{0,-\frac{1}{j}} = g_{0,j} + g_{\infty,j}$, since $T$ interchanges 0 and $\infty$ and interchanges quadratic residues and non-squares. Hence row $\infty$ of $H$ is the sum of two rows of $G$. The case $i = 0$ is treated analogously.

If $i \in Q$ then we consider the sum of the $i$-th row of $H$ and the $i$-th row of $G$. Note that $g_{ij} = g_{0,j-i}$ and $h_{ij} = g_{0,-\frac{1}{j}+\frac{1}{i}}$. This shows that $g_{ij} + h_{ij}$ is equal to zero for $j = i$ and $j = \infty$ and equal to one for $j = 0$. If $j \in N$ then we have $j - i \in Q$ if and only if $\frac{j-i}{ij} \in N$, and hence $g_{ij} + h_{ij} = 1$. If $j \in Q$ then we have $j - i \in Q$ if and only if $\frac{j-i}{ij} \in Q$, and hence $g_{ij} + h_{ij} = 0$. This proves that the sum of the $i$-th rows of $H$ and $G$ is the sum of row 0 and row $\infty$ of $G$. Hence the $i$-th row of $H$ is the sum of 3 rows of $G$.

If $i \in N$ then one can similarly show that the $i$-th row of $H$ is the sum of the $i$-th and 0-th row of $G$.

This proves Proposition 2.16. □

**Remark 2.4** The fractional linear transformation $z \mapsto bz$, $b \in \mathbb{F}_p^*$, is in $\mathrm{PSL}_2(\mathbb{F}_p)$ if and only if $b = a^2$ for some $a \in \mathbb{F}_p^*$. If $b$ is not a square in $\mathbb{F}_p$ then this transformation sends the code $C$ to the other isomorphic code which is obtained by replacing $\alpha$ by $\alpha^{-1}$. This corresponds to the fact that $\mathrm{PGL}_2(\mathbb{F}_p)$ contains $\mathrm{PSL}_2(\mathbb{F}_p)$ as a subgroup of index 2.

Note that the group $\mathrm{PSL}_2(\mathbb{F}_p)$ is *doubly transitive*, i.e., any pair of different points can be transformed into the pair $(0, \infty)$ or into any other pair of different points.

**Example 2.9** The quadratic residue code $C$ of length $p = 7$ over $\mathbb{F}_2$ is the Hamming code $H$. The extended code $\widetilde{C}$ coincides with the extended Hamming code $\widetilde{H}$. The group

$\mathrm{PSL}_2(\mathbb{F}_7)$ is isomorphic to the group $\mathrm{GL}_3(\mathbb{F}_2) = G_{168}$ which is the automorphism group of the Hamming code $H$ (cf. Sect. 1.2). The automorphism group $\mathrm{Aut}(\widetilde{C})$ acts transitively on the positions. Hence all positions are equivalent. Hence the subgroup of $\mathrm{Aut}(\widetilde{H})$ consisting of automorphisms leaving a position invariant is again $\mathrm{Aut}(H) = G_{168}$. Therefore we get

$$|\mathrm{Aut}(\widetilde{H})| = 8 \cdot 168 \, .$$

Consider the subgroup $W \subset \mathrm{Aut}(\widetilde{H})$ which consists of all $g \in \mathrm{Aut}(\widetilde{H})$ with $g(0) = 0$ and $g(\infty) = \infty$. The group $W$ can be considered as a subgroup of $\mathscr{S}_6$ and is isomorphic to the group of the rotations of a cube. Hence it is also isomorphic to $\mathscr{S}_4$ and has order 24. The subgroup $W \cap \mathrm{PSL}_2(\mathbb{F}_7)$ consists of dilations of $\mathbb{P}_1(\mathbb{F}_7)$ by squares and hence has order 3.

**Example 2.10** Let $p = 23$ and consider the binary quadratic residue code $C$ of length $p$. This is a $[23, 12]$-code. As usual, let $d$ denote the minimum distance of $C$. The BCH bound yields $d \geq 5$, by Proposition 2.15 we obtain $d \geq 7$. By Corollary 2.3 $C$ is the binary Golay code. Hence we have got another construction of the binary Golay code. Moreover, $\widetilde{C}$ is the extended binary Golay code. By Proposition 2.16, $\mathrm{PSL}_2(\mathbb{F}_{23}) \subset \mathrm{Aut}(\widetilde{C}) = M_{24}$, where $M_{24}$ is the Mathieu group which is a sporadic simple group of order

$$|M_{24}| = 2^{10} \cdot 3^3 \cdot 5 \cdot 7 \cdot 11 \cdot 23 = 244823040 \, .$$

On the other hand

$$|\mathrm{PSL}_2(\mathbb{F}_{23})| = \frac{23}{2}(23^2 - 1) = 2^3 \cdot 3 \cdot 11 \cdot 23 \, .$$

Hence again $\mathrm{PSL}_2(\mathbb{F}_{23}) \neq \mathrm{Aut}\,(\widetilde{C})$.

But E. F. Assmus and H. F. Mattson have proved the following theorem which we quote without proof (see [66, Chap. 16, §5, Theorem 13]).

**Theorem 2.10.** *If $p \equiv 7 \pmod 8$, $\frac{1}{2}(p-1)$ prime, and $p \leq 4079$ then apart from the two exceptions $p = 7$ and $p = 23$ one has*

$$\mathrm{Aut}(\widetilde{C}) = \mathrm{PSL}_2(\mathbb{F}_p) \, .$$

V. Remmert [75] has shown that the statement of Theorem 2.10 holds for all prime numbers $p = 8m - 1$ with $m > 3$.

As an application of Proposition 2.16 we prove

**Corollary 2.5.** *The minimum weight of a binary quadratic residue code $C$ is odd.*

*Proof.* Let $\widetilde{a} \neq 0$ be a word of minimum weight in $\widetilde{C}$. Since $\mathrm{PSL}_2(\mathbb{F}_p)$ is transitive on the positions and contained in $\mathrm{Aut}(\widetilde{C})$, we may assume that $\widetilde{a}$ has a 1 at the check position $\infty$. Omitting the position $\infty$ yields a word $a$ in $C$ with $w(a) = w(\widetilde{a}) - 1$. But $\widetilde{C}$ is self-dual and doubly even, and therefore all weights are even. This proves Corollary 2.5. □

Denote by $d(C)$ (resp. $d(\widetilde{C})$) the minimum weight of the code $C$ (resp. $\widetilde{C}$). Combining Proposition 2.15 and Corollary 2.5 we get

$$d(C) \equiv 3 \pmod 4,$$
$$d(C)^2 - d(C) + 1 \geq p,$$
$$d(\widetilde{C}) = d(C) + 1.$$

**Example 2.11** For $p = 31$ we get $d(C) \geq 7$ and $d(\widetilde{C}) \geq 8$. Hence $\widetilde{C}$ is an extremal doubly even self-dual code. By Example 2.5 the minimum distance of $\widetilde{C}$ must be equal to 8 and $A_8 = 620$.

**Exercise 2.6** Determine the numbers $A_7$ and $A_8$ for the binary quadratic residue code $C$ of length 31. (Clearly $A_7 + A_8 = 620$.)

**Example 2.12** For $p = 47$ we get $d(C) \geq 11$ and $d(\widetilde{C}) \geq 12$. Again $\widetilde{C}$ is an extremal doubly even self-dual code and the minimum distance is equal to 12. By Example 2.6 we have $A_{12} = 17296$.

**Example 2.13** For $p = 71$ we also derive $d(C) \geq 11$ and $d(\widetilde{C}) \geq 12$. One can show that $d(\widetilde{C}) = 12$. This means that the extended quadratic residue code $\widetilde{C}$ of length 72 is not extremal. As already mentioned, it is not known whether there exists an extremal doubly even self-dual code of length 72.

**Example 2.14** For $p = 167$ the minimum weight of $\widetilde{C}$ is unknown. One only knows that $16 \leq d(\widetilde{C}) \leq 24$.

# Chapter 3
# Even Unimodular Lattices

## 3.1 Theta Functions with Spherical Coefficients

In this section we study modified theta functions, namely theta series with spherical coefficients, and their behavior under transformations of the modular group. The results of this section are due to E. Hecke [32] and B. Schoeneberg [78, 79]. Our presentation follows [71, Chap. VI] and [83].

Let $P \in \mathbb{C}[x_1,\ldots,x_n]$ be a complex polynomial in $n$ variables $x_1,\ldots,x_n$. Recall from Sect. 1.4 that such a polynomial $P$ is called harmonic or spherical, if and only if $\Delta P = 0$, where

$$\Delta = \sum_{i=1}^{n} \frac{\partial}{\partial x_i^2}$$

is the Laplace operator. If $P$ is spherical and homogeneous of degree $r$, then $P$ is called *spherical of degree* $r$.

**Theorem 3.1.** *A polynomial* $P \in \mathbb{C}[x_1,\ldots x_n]$ *is spherical of degree* $r$ *if and only if* $P$ *is a linear combination of functions of the form* $(\xi \cdot x)^r$, *where* $\xi \in \mathbb{C}^n$, $\xi^2 = 0$ *if* $r \geq 2$.

*Proof.* (Cf. [71, Chap. VI, Proof of Theorem 18].)

"$\Leftarrow$": Let $P = (\xi \cdot x)^r$ with $\xi \in \mathbb{C}^n$, $\xi^2 = 0$. Then

$$\Delta P = \sum_i \frac{\partial}{\partial x_i^2} \left( \sum_j \xi_j x_j \right)^r = r(r-1) \left( \sum_i \xi_i^2 \right) (\xi \cdot x)^{r-2} = 0.$$

"$\Rightarrow$": For the proof of the other implication we consider the inner product on functions $f, g : \mathbb{R}^n \to \mathbb{C}$ defined by

$$(f,g) = \int_K f(x)\overline{g(x)}\,dx$$

where $K = \{x \in \mathbb{R}^n \mid \sum x_i^2 \leq 1\}$, $dx = dx_1 \ldots dx_n$. Let $P \in \mathbb{C}[x_1,\ldots,x_n]$ be a spherical polynomial of degree $r$ which is orthogonal (with respect to this inner product) to all functions of the form $(\xi \cdot x)^r$, where $\xi \in \mathbb{C}^n$ and $\xi^2 = 0$ if $r \geq 2$. We show that $P = 0$.

In order to show this, we need four formulas. Let $f$ be a homogeneous polynomial of degree $r$ in $x_1,\dots,x_n$. Then

$$\sum_i \frac{\partial f(x)}{\partial x_i}\cdot x_i = rf(x). \tag{3.1}$$

Let $\omega_i := (-1)^{i-1}dx_1\dots\widehat{dx_i}\dots dx_n$, $\omega := \sum x_i\omega_i$. Then $dx = dx_i\omega_i$. Integrating equation (3.1) and applying Stokes' theorem gives

$$r\int_{\partial K} f\omega = \int_{\partial K}\left(\sum_i \frac{\partial f}{\partial x_i}\cdot x_i\right)\omega = \int_K \Delta f(x)\,dx, \tag{3.2}$$

where $\partial K = \{x\in\mathbb{R}^n \mid \sum x_i^2 = 1\}$. A third formula is

$$\int_K \Delta f(x)dx = r(r+n)\int_K f(x)\,dx. \tag{3.3}$$

This follows from (3.2) and Stokes' theorem:

$$\int_{\partial K} f\omega = \int_{\partial K}\sum_i fx_i\omega_i = \int_K\sum_i \frac{\partial}{\partial x_i}(fx_i)\,dx = (r+n)\int_K f(x)\,dx.$$

Finally we have

$$\Delta(fg) = f\Delta g + g\Delta f + 2\sum_i \frac{\partial f}{\partial x_i}\frac{\partial g}{\partial x_i}. \tag{3.4}$$

Now let $g = (\xi\cdot x)^r$, where $\xi^2 = 0$ and $r\geq 2$. (The claim is trivial for $r = 1$.) Then $g$ and all its partial derivatives are spherical, and the same is true for $P$ and its partial derivatives. Then, from equations (3.3) and (3.4),

$$\begin{aligned}0 &= \int_K P(x)\overline{g(x)}\,dx = \text{ const.}\int_K \Delta(Pg)(x)\,dx = \text{ const.}\int_K \sum_{i_1,\dots,i_r}\frac{\partial P(x)}{\partial x_i}\frac{\partial g(x)}{\partial x_i}\,dx\\ &= \dots = \text{ const.}\int_K\sum_i \frac{\partial^r P(x)}{\partial x_{i_1}\dots\partial x_{i_r}}\frac{\partial^r g(x)}{\partial x_{i_1}\dots\partial x_{i_r}}\,dx.\end{aligned}$$

Now iteration of (3.1) gives

$$\begin{aligned}r!\,P &= \sum_{i_1,\dots,i_r}\frac{\partial^r P}{\partial x_{i_1}\dots\partial x_{i_r}}x_{i_1}\dots x_{i_r},\\ r!\,\xi_{i_1}\dots\xi_{i_r} &= \frac{\partial^r g}{\partial x_{i_1}\dots\partial x_{i_r}}.\end{aligned}$$

So the above equation yields $P(\xi) = 0$ when $\xi^2 = 0$. Hence $P(x)$ is divisible by $\delta(x) = x^2 = \sum x_i^2$. So $P(x) = \delta(x)f(x)$ for some polynomial $f(x)$. Then equations (3.2) and (3.3) yield (note that $\delta \equiv 1$ on $\partial K$)

$$\begin{aligned}\int_K f(x)\overline{f(x)}\,dx &= \text{const.}\int_{\partial K} f\bar{f}\omega \\ &= \text{const.}\int_{\partial K} \delta f\bar{f}\omega \\ &= \text{const.}\int_K P(x)\overline{f(x)}\,dx.\end{aligned}$$

We shall show that $P$ is orthogonal to all homogeneous polynomials of degree $< r$. Therefore

$$\int_K f(x)\overline{f(x)}\,dx = \text{const.}\int_K P(x)\overline{f(x)}\,dx = 0.$$

Thus $f = 0$, so $P = 0$.

It remains to show that $P$ is orthogonal to all homogeneous polynomials $f$ of degree $< r$. We show this by induction on $r$. From equations (3.3) and (3.4) and the induction hypothesis we get

$$\begin{aligned}\int_K P(x)\overline{f(x)}\,dx &= \text{const.}\int_K \Delta(Pf)(x)dx \\ &= \text{const.}\int_K P(x)\Delta f(x)\,dx \\ &= \text{const.}\int_K P(x)\Delta^2 f(x)\,dx \\ &= \dots \\ &= 0.\end{aligned}$$

This proves Theorem 3.1. □

Now let $\Gamma \subset \mathbb{R}^n$ be a lattice, $z \in \mathbb{R}^n$ be a point in $\mathbb{R}^n$, and let $P$ be a spherical polynomial of degree $r$. Let $\tau \in \mathbb{H}$. We define

**Definition.**

$$\vartheta_{z+\Gamma,P}(\tau) := \sum_{x\in z+\Gamma} P(x)e^{\pi i\tau x^2} = \sum_{x\in\Gamma} P(x+z)e^{\pi i\tau(x+z)^2}.$$

The function $\vartheta_{z+\Gamma,P}$ is holomorphic in $\mathbb{H}$. It follows from the definition that $\vartheta_{z_1+\Gamma,P} = \vartheta_{z_2+\Gamma,P}$ if $z_1, z_2 \in \mathbb{R}^n$, $z_1 \equiv z_2 \pmod{\Gamma}$. We now want to study the behavior of $\vartheta_{z+\Gamma,P}$ under substitutions of the modular group. We want to apply the Poisson summation formula. For that purpose we need to know some Fourier transforms.

**Lemma 3.1.** *We have Table 3.1 of Fourier transforms. Here $f(x)$ is the original function, and $\widehat{f}(y)$ is its Fourier transform.*

**Table 3.1** Fourier transforms

| $f(x)$ | $\widehat{f}(y)$ |
|---|---|
| $f(x)=e^{\pi i(\frac{-1}{\tau})x^2}$ $(\tau\in\mathbb{H})$ | $\widehat{f}(y)=\left(\sqrt{\frac{\tau}{i}}\right)^n e^{\pi i\tau y^2}$ $\left(-\frac{\pi}{4}<\arg\sqrt{\frac{\tau}{i}}<\frac{\pi}{4}\right)$ |
| $f(x)=e^{\pi i(\frac{-1}{\tau})(x+z)^2}$ $(\tau\in\mathbb{H},\, z\in\mathbb{R}^n)$ | $\widehat{f}(y)=\left(\sqrt{\frac{\tau}{i}}\right)^n e^{2\pi iy\cdot z}e^{\pi i\tau y^2}$ |
| $f(x)=P(x)e^{-\pi x^2}$ $(P\in\mathbb{C}[x_1,\dots,x_n])$ | $\widehat{f}(y)=P\left(\frac{-1}{2\pi i}\frac{\partial}{\partial y_1},\dots,\frac{-1}{2\pi i}\frac{\partial}{\partial y_n}\right)e^{-\pi y^2}$ |
| $f(x)=(\xi\cdot x)^r e^{-\pi x^2}$ $(\xi\in\mathbb{C}^n,\, \xi^2=0 \text{ for } r\geq 2)$ | $\widehat{f}(y)=\left(\frac{\xi\cdot y}{i}\right)^r e^{-\pi y^2}$ |
| $f(x)=(\xi\cdot(x+z))^r e^{\pi i(\frac{-1}{\tau})(x+z)^2}$ | $\widehat{f}(y)=\left(\sqrt{\frac{\tau}{i}}\right)^{n+2r}\left(\frac{\xi\cdot y}{i}\right)^r e^{2\pi iz\cdot y}e^{\pi i\tau y^2}$ |

*Proof.* For the first row see the proof of Proposition 2.1. It is equivalent to

$$\int_{\mathbb{R}^n} e^{\pi i(-\frac{1}{\tau})x^2}e^{-2\pi ix\cdot y}\,dx=\left(\sqrt{\frac{\tau}{i}}\right)^n e^{\pi i\tau y^2}. \tag{3.5}$$

The second row follows from this equality by replacing $x$ by $x+z$.
The third row follows from applying the differential operator

$$P\left(-\frac{1}{2\pi i}\frac{\partial}{\partial y_1},\dots,-\frac{1}{2\pi i}\frac{\partial}{\partial y_n}\right)$$

to the equation

$$\int_{\mathbb{R}^n} e^{-\pi x^2}e^{-2\pi ix\cdot y}\,dx=e^{-\pi y^2}.$$

(This equation is Lemma 2.1.)
The fourth row is a special case of the third row:

$$\left(-\frac{1}{2\pi i}\xi\nabla\right)^r e^{-\pi y^2}=\left(\frac{\xi\cdot y}{i}\right)^r e^{-\pi y^2},\quad \nabla:=\left(\frac{\partial}{\partial y_1},\dots,\frac{\partial}{\partial y_n}\right).$$

The fifth row follows from applying the differential operator $\left(-\frac{1}{2\pi i}\xi\nabla\right)^r$ to equation (3.5) and then replacing $x$ by $x+z$. □

**Proposition 3.1.** *We have the identity*

$$\vartheta_{z+\Gamma,P}\left(-\frac{1}{\tau}\right)=\left(\sqrt{\frac{\tau}{i}}\right)^{n+2r} i^{-r}\frac{1}{\operatorname{vol}(\mathbb{R}^n/\Gamma)}\sum_{y\in\Gamma^*}P(y)e^{2\pi iy\cdot z}e^{\pi i\tau y^2}.$$

*Proof.* By Theorem 3.1 $\vartheta_{z+\Gamma,P}(\tau)$ is a finite sum of series of the form

$$\sum_{x\in\Gamma}(\xi\cdot(x+z))^r e^{\pi i\tau(x+z)^2},\qquad \xi\in\mathbb{C}^n,\ \xi^2=0 \text{ for } r\geq 2.$$

Now by Lemma 3.1 and the Poisson summation formula (Theorem 2.3)

$$\sum_{x\in\Gamma} (\xi\cdot(x+z))^r e^{\pi i\left(-\frac{1}{\tau}\right)(x+z)^2}$$
$$= \left(\sqrt{\frac{\tau}{i}}\right)^{n+2r} i^{-r}\frac{1}{\operatorname{vol}(\mathbb{R}^n/\Gamma)}\sum_{y\in\Gamma^*}(\xi\cdot y)^r e^{2\pi i y\cdot z}e^{\pi i\tau y^2}.$$

This gives the formula of Proposition 3.1. □

For the remaining part of this section we make the *general assumption*, that $\Gamma\subset\mathbb{R}^n$ is an even integral lattice, that $n$ is even, that $P$ is spherical of degree $r$, and that $k := \frac{n}{2}+r$.

Note that $\Gamma\subset\Gamma^*$, and that $y_1, y_2\in\Gamma^*$, $y_1\equiv y_2 \pmod{\Gamma}$ implies that $y_1\cdot x\equiv y_2\cdot x \pmod{\mathbb{Z}}$ for all $x\in\Gamma$ and $y_1^2\equiv y_2^2 \pmod{2\mathbb{Z}}$. Therefore we have the following formulas for $\rho\in\Gamma^*$ ($v(\Gamma) := \operatorname{vol}(\mathbb{R}^n/\Gamma)$):

(T1) $$\vartheta_{\rho+\Gamma,P}(\tau+1) = e^{\pi i\rho^2}\vartheta_{\rho+\Gamma,P}(\tau),$$

(T2) $$\vartheta_{\rho+\Gamma,P}\left(-\frac{1}{\tau}\right) = \frac{1}{v(\Gamma)}\left(\frac{\tau}{i}\right)^k i^{-r}\sum_{\sigma\in\Gamma^*/\Gamma} e^{2\pi i\sigma\rho}\vartheta_{\sigma+\Gamma,P}(\tau).$$

The group $SL_2(\mathbb{Z})$ operates on $\mathbb{H}$ and on the set of all functions $f:\mathbb{H}\to\mathbb{C}$ by associating to $f$ the function $f|_kA$ defined by

$$(f|_kA)(\tau) := f(A\tau)(c\tau+d)^{-k}$$

for $\tau\in\mathbb{H}$, $A\in SL_2(\mathbb{Z})$. The formulas (T1) and (T2) describe the transformation of $\vartheta_{\rho+\Gamma,P}$ by $T$ and $S$ respectively, where $S$ and $T$ are the transformations of Sect. 2.2 which generate $SL_2(\mathbb{Z})$.

**Proposition 3.2.** *The group* $SL_2(\mathbb{Z})$ *leaves the span of all* $\vartheta_{\rho+\Gamma,P}$ $(\rho\in\Gamma^*)$ *invariant. More precisely we have the following formulas. Let* $\rho\in\Gamma^*$, $A=\begin{pmatrix} a & b\\ c & d\end{pmatrix}\in SL_2(\mathbb{Z})$. *Then we have*

$$\vartheta_{\rho+\Gamma,P}\big|_k A = \frac{1}{v(\Gamma)c^{n/2}i^{k+r}}$$
$$\cdot\sum_{\sigma\in\Gamma^*/\Gamma}\left(e^{-\pi i b(d\sigma^2+2\sigma\rho)}\sum_{\substack{\lambda\in\Gamma^*/c\Gamma\\ \lambda\equiv\rho+d\sigma\,(\Gamma)}} e^{\pi i\frac{a}{c}\lambda^2}\right)\vartheta_{\sigma+\Gamma,P}$$

*if* $c\neq 0$, *and*

$$\vartheta_{\rho+\Gamma,P}\big|_k A = \frac{1}{d^{n/2}}e^{\pi i a b\rho^2}\vartheta_{a\rho+\Gamma,P}$$

*if* $c = 0$.

**Remark 3.1** One has

$$\dim_{\mathbb{C}} \operatorname{span} \left\{ \vartheta_{\rho+\Gamma,P} \mid \rho \in \Gamma^*/\Gamma \right\} \leq |\Gamma^*/\Gamma|.$$

In general the left-hand side is smaller than the right-hand side, because e.g. $\vartheta_{\Gamma,P} = 0$, if $r = \deg P$ is odd, or

$$\vartheta_{\rho+\Gamma,P} = (-1)^r \vartheta_{-\rho+\Gamma,P}.$$

*Proof of Proposition 3.2.* The formula for $c = 0$ is a generalization of the formula (T1) and follows from the definition of $\vartheta_{\rho+\Gamma}$:

$$\begin{aligned}
\vartheta_{\rho+\Gamma,P}\left(a^2\tau + ab\right) &= \sum_{x\in\rho+\Gamma} P(x) e^{\pi i\left(a^2\tau+ab\right)x^2} \\
&= e^{\pi i a b \rho^2} \sum_{x\in\rho+\Gamma} P(x) e^{\pi i \tau (ax)^2} \\
&= e^{\pi i a b \rho^2} a^{-r} \sum_{x\in a\rho+\Gamma} P(x) e^{\pi i \tau x^2} \\
&= d^r e^{\pi i a b \rho^2} \vartheta_{a\rho+\Gamma,P}(\tau).
\end{aligned}$$

We now assume that $c \neq 0$. We may assume without loss of generality that $c > 0$, since

$$\vartheta_{\rho+\Gamma,P}\big|_k A = \vartheta_{\rho+\Gamma,P}\big|_k \left(\begin{pmatrix} -1 & 0 \\ 0 & -1 \end{pmatrix}\right)\Big|_k \begin{pmatrix} -a & -b \\ -c & -d \end{pmatrix}.$$

We use a little device which is sometimes attributed to Hermite. We write

$$\frac{a\tau+b}{c\tau+d} = \frac{a}{c} - \frac{1}{c(c\tau+d)}.$$

Since $y_1, y_2 \in \Gamma^*$, $y_1 \equiv y_2 \pmod{c\Gamma}$ implies that $y_1^2 \equiv y_2^2 \pmod{2c\mathbb{Z}}$, we can write

$$\vartheta_{\rho+\Gamma,P} = \sum_{\substack{\lambda\in\Gamma^*/c\Gamma \\ \lambda\equiv\rho\,(\Gamma)}} \vartheta_{\lambda+c\Gamma,P}.$$

Now by applying the formulas (T1) and (T2), we get

$$\begin{aligned}
&\vartheta_{\lambda+c\Gamma,P}\left(\frac{a}{c} - \frac{1}{c(c\tau+d)}\right) \\
&= e^{\pi i \frac{a}{c}\lambda^2} \vartheta_{\lambda+c\Gamma,P}\left(\frac{-1}{c(c\tau+d)}\right) \\
&= \frac{c^k}{v(c\Gamma) i^{k+r}} (c\tau+d)^k \sum_{\sigma\in(c\Gamma)^*/c\Gamma} e^{\pi i\left(\frac{a}{c}\lambda^2+2\sigma\lambda\right)} \vartheta_{\sigma+c\Gamma,P}\left(c^2\tau+cd\right)
\end{aligned}$$

$$= \frac{c^k}{v(c\Gamma)i^{k+r}}(c\tau+d)^k \sum_{\sigma\in(c\Gamma)^*/c\Gamma} e^{\pi i\left(\frac{a}{c}\lambda^2+2\lambda\sigma+cd\sigma^2\right)}\vartheta_{\sigma+c\Gamma,P}\left(c^2\tau\right).$$

We now sum over $\frac{\sigma}{c}$ instead of $\sigma$. Note that $(c\Gamma)^* = \frac{1}{c}\Gamma^*$ and $v(c\Gamma) = c^n v(\Gamma)$. Thus

$$\begin{aligned}\left(\vartheta_{\lambda+c\Gamma,P}\big|_k A\right)(\tau) &= \frac{c^r}{v(\Gamma)c^{n/2}\,i^{k+r}} \sum_{\sigma\in\Gamma^*/c^2\Gamma} e^{\pi i\left(\frac{a}{c}\lambda^2+2\lambda\frac{\sigma}{c}+\frac{d}{c}\sigma^2\right)}\vartheta_{\frac{\sigma}{c}+c\Gamma,P}(c^2\tau)\\ &= \frac{1}{v(\Gamma)c^{n/2}\,i^{k+r}} \sum_{\sigma\in\Gamma^*/c^2\Gamma} e^{\pi i\left(\frac{a}{c}\lambda^2+2\lambda\frac{\sigma}{c}+\frac{d}{c}\sigma^2\right)}\vartheta_{\sigma+c^2\Gamma,P}(\tau).\end{aligned}$$

Therefore we get

$$\left(\vartheta_{\rho+\Gamma,P}\big|_k A\right)(\tau) = \sum_{\sigma\in\Gamma^*/c^2\Gamma} G(\sigma)\vartheta_{\sigma+c^2\Gamma,P}(\tau)$$

where

$$G(\sigma) := \frac{1}{v(\Gamma)c^{n/2}\,i^{k+r}} \sum_{\substack{\lambda\in\Gamma^*/c\Gamma\\ \lambda\equiv\rho\,(\Gamma)}} e^{\frac{\pi i}{c}\left(a\lambda^2+2\lambda\sigma+d\sigma^2\right)}.$$

Now

$$\begin{aligned}\frac{a}{c}\lambda^2+\frac{2\lambda\sigma}{c}+\frac{d}{c}\sigma^2 &= \frac{a}{c}\lambda^2+\frac{2(ad-bc)}{c}\lambda\sigma+\frac{d}{c}\sigma^2\\ &= \frac{a}{c}(\lambda+d\sigma)^2-\frac{ad^2}{c}\sigma^2-2b\lambda\sigma+\frac{d}{c}\sigma^2\\ &= \frac{a}{c}(\lambda+d\sigma)^2-2b\lambda\sigma+d\frac{1-ad}{c}\sigma^2.\end{aligned}$$

Thus

$$\begin{aligned}G(\sigma) &= \frac{1}{v(\Gamma)c^{n/2}\,i^{k+r}}e^{-\pi ib\left(d\sigma^2+2\sigma\rho\right)} \sum_{\substack{\lambda\in\Gamma^*/c\Gamma\\ \lambda\equiv\rho\,(\Gamma)}} e^{\pi i\frac{a}{c}(\lambda+d\sigma)^2}\\ &= \frac{1}{v(\Gamma)c^{n/2}\,i^{k+r}}e^{-\pi ib\left(d\sigma^2+2\sigma\rho\right)} \sum_{\substack{\lambda\in\Gamma^*/c\Gamma\\ \lambda\equiv\rho+d\sigma\,(\Gamma)}} e^{\pi i\frac{a}{c}\lambda^2}.\end{aligned}$$

This shows in particular that the coefficient $G(\sigma)$ only depends on $\sigma \bmod \Gamma$. Therefore we have

$$\left(\vartheta_{\rho+\Gamma,P}\big|_k A\right)(\tau) = \sum_{\sigma\in\Gamma^*/\Gamma} G(\sigma)\vartheta_{\sigma+\Gamma,P}(\tau),$$

yielding the formula of Proposition 3.2 for $c\neq 0$. □

We examine the coefficients of the formula of Proposition 3.2 for $c\neq 0$ more closely. Let

$$S := \sum_{\substack{\lambda \in \Gamma^*/c\Gamma \\ \lambda \equiv \rho + d\sigma\,(\Gamma)}} e^{\pi i \frac{a}{c} \lambda^2}.$$

Replacing $\lambda$ by $\lambda + c\mu$, where $\mu \in \Gamma^*$, $c\mu \in \Gamma$, we get

$$\begin{aligned} S &= \sum_{\substack{\lambda \in \Gamma^*/c\Gamma \\ \lambda \equiv \rho + d\sigma(\Gamma)}} e^{\pi i \frac{a}{c} (\lambda + c\mu)^2} \\ &= e^{\pi i a c \mu^2} \sum_{\lambda \ldots} e^{\pi i (\frac{a}{c} \lambda^2 + 2a\lambda\mu)} \\ &= e^{\pi i \left(ac\mu^2 + 2a(\rho + d\sigma)\mu\right)} \sum_{\lambda \ldots} e^{\pi i \frac{a}{c} \lambda^2} \\ &= e^{\pi i \left(ac\mu^2 + 2a(\rho + d\sigma)\mu\right)} \cdot S. \end{aligned}$$

Hence $S$ can only be different from zero if

$$e^{\pi i \left(ac\mu^2 + 2a(\rho + d\sigma)\mu\right)} = 1$$

for all $\mu \in \Gamma^*$ with $c\mu \in \Gamma$.

**Definition.** The minimum of all $N \in \mathbb{N}$ with $N\mu^2 \in 2\mathbb{Z}$ for all $\mu \in \Gamma^*$ is called the *level of* $\Gamma$.

**Lemma 3.2.** *Let $N$ be the level of $\Gamma$. Then $N\Gamma^* \subset \Gamma$.*

*Proof.* Let $(e_1, \ldots, e_n)$ be a basis of $\Gamma$, and let $A$ be the matrix $((e_i \cdot e_j))$ (cf. Sect. 1.1). We claim that $NA^{-1}$ is an integral matrix. In fact, let $(e_1^*, \ldots, e_n^*)$ be the dual basis of $(e_1, \ldots, e_n)$. Now

$$N(e_i^* \cdot e_j^*) = \frac{1}{2} N \left( (e_i^* + e_j^*)^2 - (e_i^*)^2 - (e_j^*)^2 \right),$$

and by definition of $N$, $N\mu^2 \in 2\mathbb{Z}$ for all $\mu \in \Gamma^*$. Hence $Ne_i^* \cdot e_j^* \in \mathbb{Z}$ for all $1 \leq i,\ j \leq n$, and therefore $NA^{-1} = \left((Ne_i^* \cdot e_j^*)\right)$ (cf. Sect. 1.1) is an integral matrix. But

$$Ne_i^* = \sum_{j=1}^{n} N b_{ij} e_j,$$

where $B = ((b_{ij})) = A^{-1}$. Since $NB = NA^{-1}$ is an integral matrix, $Ne_i^* \in \Gamma$ for all $1 \leq i \leq n$. This proves $N\Gamma^* \subset \Gamma$, and hence the lemma. □

**Remark 3.2** Let $\Delta = (-1)^{n/2} \operatorname{disc}(\Gamma)$ and let $N$ be the level of $\Gamma$. Then $\Delta A^{-1}$ is an integral matrix with even diagonal elements. The same is true for $NA^{-1}$ as we have shown in the proof of Lemma 3.2. It follows that $N \mid \Delta$ and $\Delta \mid N^n$.

Now let $N$ be the level of $\Gamma$ and suppose that $N \mid c$. Then $S$ can only be different from zero if $a(\rho + d\sigma)\mu \in \mathbb{Z}$ for all $\mu \in \Gamma^*$. But this is equivalent to $a(\rho + d\sigma) \in \Gamma^{**} = \Gamma$. But then $a\rho + \sigma \in \Gamma$ since $bc\sigma \in \Gamma$ by Lemma 3.2. Finally we have

$$\vartheta_{a\rho+\Gamma,P} = (-1)^r \vartheta_{-a\rho+\Gamma,P}.$$

Therefore we have proved

**Corollary 3.1.** *Let $\Gamma \subset \mathbb{R}^n$ (n even) be an even integral lattice of level $N$, $\rho \in \Gamma^*$, $A = \begin{pmatrix} a & b \\ c & d \end{pmatrix} \in \mathrm{SL}_2(\mathbb{Z})$, and suppose $N \mid c$. Then*

$$\vartheta_{\rho+\Gamma,P}\big|_k A = \varepsilon(A) e^{\pi i a b \rho^2} \vartheta_{a\rho+\Gamma,P},$$

*where*

$$\varepsilon(A) = \begin{cases} \frac{1}{v(\Gamma)(ic)^{n/2}} \sum_{\lambda \in \Gamma/c\Gamma} e^{\pi i \frac{a}{c} \lambda^2} & \text{for } c \neq 0, \\ d^{-n/2} & \text{for } c = 0. \end{cases}$$

One can consider the following subgroups of finite index of $\mathrm{SL}_2(\mathbb{Z})$:

$$\Gamma_0(N) := \left\{ \begin{pmatrix} a & b \\ c & d \end{pmatrix} \in \mathrm{SL}_2(\mathbb{Z}) \,\middle|\, N \mid c \right\}$$

$$\Gamma_1(N) := \left\{ \begin{pmatrix} a & b \\ c & d \end{pmatrix} \in \mathrm{SL}_2(\mathbb{Z}) \,\middle|\, a \equiv d \equiv 1\ (N),\ c \equiv 0\ (N) \right\}$$

$$\Gamma(N) := \left\{ \begin{pmatrix} a & b \\ c & d \end{pmatrix} \in \mathrm{SL}_2(\mathbb{Z}) \,\middle|\, \begin{pmatrix} a & b \\ c & d \end{pmatrix} \equiv \begin{pmatrix} 1 & 0 \\ 0 & 1 \end{pmatrix} (N) \right\}.$$

The last group is the principal congruence subgroup of level $N$ of $\mathrm{SL}_2(\mathbb{Z})$. It is a normal subgroup of $\mathrm{SL}_2(\mathbb{Z})$.

By Corollary 3.1 one has in particular

$$\vartheta_{\Gamma,P}\big|_k A = \varepsilon(A) \vartheta_{\Gamma,P}$$

for $A \in \Gamma_0(N)$. This formula implies that the mapping $\varepsilon : \Gamma_0(N) \to \mathbb{C}^*$ is a group homomorphism. Such a group homomorphism is called a *character* of $\Gamma_0(N)$. We want to determine this character. For the following corollary cf. [28, §22, Corollary to Lemma 9].

**Corollary 3.2 (Reciprocity law for Gaussian sums).** *Let $A \in \Gamma_0(N)$, $A = \begin{pmatrix} a & b \\ c & d \end{pmatrix}$, $d, c \neq 0$. Then*

$$\varepsilon(A) = d^{-n/2} \sum_{\lambda \in \Gamma/d\Gamma} e^{\pi i \frac{b}{d} \lambda^2}.$$

*Proof.* We write

$$\vartheta_{\Gamma,P}\big|_k A = \vartheta_{\Gamma,P}\big|_k \left( A \cdot \begin{pmatrix} 0 & -1 \\ 1 & 0 \end{pmatrix}^{-1} \right)\Bigg|_k \begin{pmatrix} 0 & -1 \\ 1 & 0 \end{pmatrix},$$

and apply twice the transformation formula of Proposition 3.2. □

**Lemma 3.3.** *Let $A \in \Gamma_0(N)$, $A = \begin{pmatrix} a & b \\ c & d \end{pmatrix}$. Then*

$$\varepsilon \begin{pmatrix} a & b \\ c & d \end{pmatrix} = \varepsilon \begin{pmatrix} a & b+la \\ c & d+lc \end{pmatrix}$$

*for any integer $l$.*

*Proof.* This follows directly from Corollary 3.1. □

Let $G(b,d)$ (for $d \neq 0$) be the Gaussian sum

$$G(b,d) := \sum_{\lambda \in \Gamma / d\Gamma} e^{\pi i \frac{b}{d} \lambda^2}.$$

For the following lemma cf. [28, §22, Lemma 10].

**Lemma 3.4.** *Let $A \in \Gamma_0(N)$, $A = \begin{pmatrix} a & b \\ c & d \end{pmatrix}$, $d, c \neq 0$. Then the Gaussian sum $G(b,d)$ is a rational number. Moreover, $G(b,d) = G(1,d)$.*

*Proof.* Since $\Gamma$ is even, $G(b,d)$ is a sum of $d^{\text{th}}$ roots of unity and hence lies in the field $\mathbb{Q}(\zeta)$, where $\zeta = e^{2\pi i/d}$. By Corollary 3.2,

$$G(b,d) = d^{n/2} \varepsilon(A).$$

By Lemma 3.3 we get

$$G(b,d) = d^{n/2} \varepsilon(A) = d^{n/2} (d+lc)^{-n/2} \sum_{\lambda \in \Gamma / (d+lc)\Gamma} e^{\pi i \frac{b+la}{d+lc} \lambda^2}$$

for all integers $l$. Hence $G(b,d) \in \mathbb{Q}(\zeta_l)$, where $\zeta_l = e^{2\pi i/(d+lc)}$, for all integers $l$. Since $c$ and $d$ are coprime, there exists an $l$ such that $d$ and $d+lc$ are coprime. For this $l$ we have $\mathbb{Q}(\zeta) \cap \mathbb{Q}(\zeta_l) = \mathbb{Q}$, thus $G(b,d) \in \mathbb{Q}$. Moreover, $G(b,d)$ is invariant under all automorphisms of $\mathbb{Q}(\zeta)$. Using the automorphism $\zeta \mapsto \zeta^b$, we see that $G(b,d) = G(1,d)$. This proves Lemma 3.4. □

Let $A \in \Gamma_0(N)$, $A = \begin{pmatrix} a & b \\ c & d \end{pmatrix}$. Corollary 3.2 and Lemma 3.4 show that $\varepsilon(A)$ depends only on $d$. In fact, Lemma 3.3 implies that $\varepsilon(A)$ depends only on the congruence class of $d \bmod N$. Therefore, using the formula of Corollary 3.2, we get $\varepsilon(A) = 1$ for $A \in \Gamma_1(N)$. Therefore $\varepsilon$ factors to a rational character of $\Gamma_0(N)/\Gamma_1(N)$, and hence, via $\Gamma_0(N)/\Gamma_1(N) \xrightarrow{\cong} (\mathbb{Z}/N\mathbb{Z})^*$, $A \longmapsto d \bmod N$, to a character of $(\mathbb{Z}/N\mathbb{Z})^*$, which we denote by $\chi$.

It remains to determine $\chi(d)$ $(= \chi(d \bmod N))$ for an integer $d$. It suffices to compute $\chi(p)$ for a prime number $p$, $p \neq 2$, which is coprime to $N$. Choose integers $t$ and $u$ with $pt - uN = 1$. Now we apply Corollary 3.1 to

$$A = \begin{pmatrix} p & 1 \\ uN & t \end{pmatrix}.$$

Since $\chi(t) = \chi(p)$, we get

$$\begin{aligned}
\chi(p)v(\Gamma)(iuN)^{n/2} &= \sum_{\lambda \in \Gamma/uN\Gamma} e^{\pi i \frac{p}{uN}\lambda^2} \\
&\equiv \left( \sum_{\lambda \in \Gamma/uN\Gamma} e^{\pi i \frac{\lambda^2}{uN}} \right)^p \pmod p \\
&\equiv \left( \varepsilon\left(\begin{pmatrix} 1 & 1 \\ uN & pt \end{pmatrix}\right) v(\Gamma)(iuN)^{n/2} \right)^p \pmod p.
\end{aligned}$$

Now $\varepsilon\left(\begin{pmatrix} 1 & 1 \\ uN & pt \end{pmatrix}\right) = 1$, and $v(\Gamma) = \operatorname{disc}(\Gamma)^{1/2}$, where $\operatorname{disc}(\Gamma)$ is the discriminant of $\Gamma$ (cf. Sect. 1.1). Moreover $p \nmid \operatorname{disc}(\Gamma)$, and thus

$$\begin{aligned}
\chi(p) &\equiv \left( v(\Gamma)(iuN)^{n/2} \right)^{p-1} \pmod p \\
&\equiv \left( \operatorname{disc}(\Gamma)(-1)^{n/2} \right)^{\frac{p-1}{2}} \pmod p.
\end{aligned}$$

If we set $\Delta = (-1)^{n/2} \operatorname{disc}(\Gamma)$, then we get

$$\chi(p) \equiv \Delta^{\frac{p-1}{2}} \pmod p \equiv \left( \frac{\Delta}{p} \right) \quad \text{( Legendre symbol )}.$$

According to common use in number theory we write

$$\left( \frac{\Delta}{d} \right)$$

instead of $\chi(d)$. So we have finally proved the following theorem.

**Theorem 3.2.** *We have*

$$\begin{aligned}
\vartheta_{\rho+\Gamma,P}\big|_k A &= \vartheta_{\rho+\Gamma,P} \qquad \text{for } A \in \Gamma(N), \\
\vartheta_{\Gamma,P}\big|_k A &= \left( \frac{\Delta}{d} \right) \vartheta_{\Gamma,P} \qquad \text{for } A \in \Gamma_0(N).
\end{aligned}$$

## 3.2 Root Systems in Even Unimodular Lattices

Now we specialize to the case where $\Gamma \subset \mathbb{R}^n$ is an even unimodular lattice. Then we have the following corollary of Theorem 3.2.

**Corollary 3.3.** *Let $\Gamma \subset \mathbb{R}^n$ be an even unimodular lattice, and let $P$ be a spherical polynomial in $n$ variables of degree $r$. Then $\vartheta_{\Gamma,P}$ is a modular form of weight $\frac{n}{2}+r$, and a cusp form if $r>0$.*

*Proof.* The first assertion follows from Theorem 3.2, since a unimodular lattice has level and discriminant 1, and $n \equiv 0 \pmod 8$. So it remains to show that $\vartheta_{\Gamma,P}$ is a cusp form if $r>0$. Now

$$\vartheta_{\Gamma,P}(\tau) = \sum_{s=0}^{\infty} c_s q^s$$

where $q = e^{2\pi i \tau}$, and

$$c_s := \sum_{\substack{x\in\Gamma \\ x^2=2s}} P(x).$$

Now if $r>0$ then $c_0 = P(0) = 0$, so $\vartheta_{\Gamma,P}$ is a cusp form. This proves Corollary 3.3. □

We want to apply Corollary 3.3 to the classification of even unimodular lattices of dimension $\leq 24$. The idea of this application is due to B. B. Venkov ([89], see also [21, Chap. 18]), and we shall follow his presentation.

We want to classify root systems in even unimodular lattices. Let

$$\Gamma_2 := \{x \in \Gamma \mid x^2 = 2\}$$

be the set of roots in $\Gamma$. Let

$$f(x) := (x\cdot y)^2 - \frac{(x\cdot x)(y\cdot y)}{n}$$

for a fixed $y \in \mathbb{R}^n$. Then the polynomial $f$ is harmonic (cf. Sect. 1.4). By Corollary 3.3, $\vartheta_{\Gamma,f}$ is a cusp form of weight $\frac{n}{2}+2$. So for $n = 8, 16, 24$ the theta function $\vartheta_{\Gamma,f}$ is a cusp form of weight 6, 10, 14 respectively. But by Theorem 2.5, such a form is identically equal to zero. So all the coefficients $c_r$ of the expansion

$$\vartheta_{\Gamma,f}(\tau) = \sum_{r=0}^{\infty} c_r q^r$$

have to be equal to zero for $n = 8, 16, 24$. So we have proved:

**Proposition 3.3.** *Let $\Gamma \subset \mathbb{R}^n$ be an even unimodular lattice, and let $n = 8, 16,$ or $24$. Then for a fixed $y \in \mathbb{R}^n$ we have*

$$\sum_{\substack{x\in\Gamma \\ x^2=2r}} (x\cdot y)^2 - \left( \sum_{\substack{x\in\Gamma \\ x^2=2r}} x^2 \right) \frac{y^2}{n} = 0.$$

*In particular we have the following equality for the roots in $\Gamma$:*

$$\sum_{x\in\Gamma_2} (x\cdot y)^2 = \frac{1}{n}\cdot 2 \cdot |\Gamma_2| \cdot y^2. \tag{3.6}$$

**Corollary 3.4.** *Either $\Gamma_2 = \emptyset$ or $\Gamma_2$ spans $\mathbb{R}^n$.*

*Proof.* If $\Gamma_2$ does not span $\mathbb{R}^n$, then there is a $y \in \mathbb{R}^n$, $y \neq 0$, which is orthogonal to all elements of $\Gamma_2$. Then Formula (3.6) implies $|\Gamma_2| = 0$. □

**Corollary 3.5.** *Let $\Gamma \subset \mathbb{R}^n$ be an even unimodular lattice, and let $n = 8, 16$, or $24$. Then all irreducible components of the root lattice spanned by $\Gamma_2$ have the same Coxeter number $h$, and $h = \frac{1}{n}|\Gamma_2|$.*

*Proof.* This corollary follows from Propositions 3.3 and 1.6 by choosing $y \in \mathbb{R}^n$ to lie in an irreducible component of the root lattice spanned by $\Gamma_2$. □

Now let us consider the implications of Corollaries 3.4 and 3.5 for the cases $n = 8$, 16, and 24 separately. We denote by $(\Gamma_2)_{\mathbb{Z}}$ the root lattice spanned by $\Gamma_2$. For $n = 8$ we get $(\Gamma_2)_{\mathbb{Z}} = E_8$ (cf. Proposition 2.5).

Let $n = 16$ and let $\Gamma \subset \mathbb{R}^{16}$ be an even unimodular lattice. Then $\vartheta_\Gamma$ is a modular form of weight 8 (by Theorem 2.1). By Corollary 2.2, $\vartheta_\Gamma = E_4^2$, where $E_4$ is the Eisenstein series of weight 4. Hence $|\Gamma_2| = 480$ by Sect. 2.5. So by Corollary 3.5 all irreducible components of $(\Gamma_2)_{\mathbb{Z}}$ have the Coxeter number 30. By Corollary 3.4 $\Gamma_2$ spans $\mathbb{R}^{16}$. A look at Table 1.1 yields the following two possibilities for $(\Gamma_2)_{\mathbb{Z}}$:

$$(\Gamma_2)_{\mathbb{Z}} = E_8 \perp E_8 \qquad \text{or} \qquad (\Gamma_2)_{\mathbb{Z}} = D_{16}.$$

**Proposition 3.4.** *Let $\Gamma \subset \mathbb{R}^{24}$ be an even unimodular lattice. Then $(\Gamma_2)_{\mathbb{Z}}$ is one of the following 24 lattices :*

$$\begin{gathered} 0, \\ 24A_1,\, 12A_2,\, 8A_3,\, 6A_4,\, 4A_6,\, 3A_8,\, 2A_{12},\, A_{24}, \\ 6D_4,\, 4D_6,\, 3D_8,\, 2D_{12},\, D_{24}, \\ 4E_6,\, 3E_8, \\ 4A_5 \perp D_4,\, 2A_7 \perp 2D_5,\, 2A_9 \perp D_6,\, A_{15} \perp D_9, \\ E_8 \perp D_{16},\, 2E_7 \perp D_{10},\, E_7 \perp A_{17},\, E_6 \perp D_7 \perp A_{11}. \end{gathered}$$

*Proof.* The proof is the combinatorial verification that these 24 possibilities for $(\Gamma_2)_{\mathbb{Z}}$ are the only possibilities for root systems $\Gamma_2$ which satisfy the following two properties:

(i) $\Gamma_2$ spans $\mathbb{R}^{24}$.
(ii) All the irreducible components of $(\Gamma_2)_{\mathbb{Z}}$ have the same Coxeter number.

Let

$$(\Gamma_2)_{\mathbb{Z}} = \bigoplus_{i=1}^{24} \alpha_i A_i \perp \bigoplus_{j=1}^{24} \beta_j D_j \perp \bigoplus_{k=6}^{8} \gamma_k E_k.$$

Condition (i) means that

$$\sum_{i=1}^{24} i\alpha_i + \sum_{j=1}^{24} j\beta_j + \sum_{k=6}^{8} k\gamma_k = 24.$$

Using Table 1.1, one easily checks that condition (ii) leads to the following equations:

$$\begin{aligned} i\alpha_i &= 24, \\ j\beta_j &= 24, \\ k\gamma_k &= 24, \\ (2j-3)\alpha_{2j-3} + j\beta_j &= 24, \\ 11\alpha_{11} + 7\beta_7 + 6\gamma_6 &= 24, \\ 17\alpha_{17} + 10\beta_{10} + 7\gamma_7 &= 24, \\ 16\beta_{16} + 8\gamma_8 &= 24. \end{aligned}$$

The solutions to these seven equations yield exactly the possibilities of Proposition 3.4. For more details see [89]. This proves Proposition 3.4. □

**Exercise 3.1** Let $\Gamma \subset \mathbb{R}^{32}$ be an even unimodular lattice. Show that for a fixed $y \in \mathbb{R}^{32}$ we have

$$-528 \sum_{\substack{x\in\Gamma \\ x^2=2}} \left( (x\cdot y)^2 - \frac{x^2y^2}{32} \right) = \sum_{\substack{x\in\Gamma \\ x^2=4}} \left( (x\cdot y)^2 - \frac{x^2y^2}{32} \right). \tag{3.7}$$

(Hint: A cusp form of weight 18 must be a multiple of the cusp form $\Delta E_6$.)

**Exercise 3.2** Let $C \subset \mathbb{F}_2^{32}$ be a doubly even self-dual code. We assume that $C$ is extremal, i.e., $A_4 = 0$ and hence $A_8 = 620$. Let $\widetilde{A}_m$ be the number of codewords of weight $m+1$ in $C$ which have a 1 at the first position. Show that $\widetilde{A}_7 = \frac{1}{4}A_8$. (Hint: Apply Formula (3.7) to the lattice $\Gamma_C \subset \mathbb{R}^{32}$ and to the vector $y = (\sqrt{2}, 0, \ldots, 0) \in \mathbb{R}^{32}$.)

**Exercise 3.3** Let $C \subset \mathbb{F}_2^{32}$ be a doubly even self-dual code, not necessarily extremal. Derive a relation between $A_4$, $\widetilde{A}_3$, $A_8$, and $\widetilde{A}_7$ from Formula (3.7).

In the next section we shall consider the question, in how many ways each of the above root lattices can be realized as a root sublattice of an even unimodular lattice of the same dimension.

## 3.3 Overlattices and Codes

We now want to discuss the existence and uniqueness of even unimodular lattices containing a given root lattice of the same dimension as the root sublattice. More generally we are concerned with the problem of the existence and uniqueness of embeddings of one lattice into another of the same dimension. This problem can be treated with the technique of discriminant forms which we now want to describe (see also [69]).

Let $\Gamma$ be an even lattice in $\mathbb{R}^n$. Then we have a canonical embedding $\Gamma \hookrightarrow \Gamma^*$ into the dual lattice of $\Gamma$. The quotient group

$$G_\Gamma := \Gamma^*/\Gamma$$

is a finite abelian group of order $\operatorname{disc}(\Gamma)$ (see Sect. 1.1). We define a mapping $b_\Gamma : G_\Gamma \times G_\Gamma \to \mathbb{Q}/\mathbb{Z}$ by

$$b_\Gamma(x+\Gamma, y+\Gamma) = x \cdot y + \mathbb{Z}, \qquad \text{where } x, y \in \Gamma^*,$$

and a mapping $q_\Gamma : G_\Gamma \to \mathbb{Q}/2\mathbb{Z}$ by

$$q_\Gamma(x+\Gamma) = x^2 + 2\mathbb{Z}, \qquad \text{where } x \in \Gamma^*.$$

Then $b_\Gamma$ is a finite symmetric bilinear form, and $q_\Gamma$ is a *finite quadratic form*. By this we mean a mapping $q : G \to \mathbb{Q}/2\mathbb{Z}$ defined on a finite abelian group $G$ satisfying the following conditions:

(i) $q(rx) = r^2 q(x)$ for all $r \in \mathbb{Z}$ and $x \in G$,
(ii) $q(x+y) - q(x) - q(y) \equiv 2b(x,y) \pmod{2\mathbb{Z}}$,

where $b : G \times G \to \mathbb{Q}/\mathbb{Z}$ is a finite symmetric bilinear form, which we call the bilinear form corresponding to $q$. The form $q_\Gamma$ is called the *discriminant quadratic form* of $\Gamma$.

Let $\Lambda$ be an even lattice in $\mathbb{R}^n$. We call an embedding $\Lambda \hookrightarrow \Gamma$, where $\Gamma$ is another even lattice in $\mathbb{R}^n$, an *even overlattice* of $\Lambda$. We are interested in classifying such overlattices of the given lattice $\Lambda$.

Let $\Lambda \hookrightarrow \Gamma$ be an even overlattice of $\Lambda$. Then we can consider the quotient group

$$H_\Gamma := \Gamma/\Lambda,$$

which is a finite abelian group of order $[\Gamma : \Lambda]$, where $[\Gamma : \Lambda]$ is the index of $\Lambda$ in $\Gamma$. We have a chain of embeddings

$$\Lambda \hookrightarrow \Gamma \hookrightarrow \Gamma^* \hookrightarrow \Lambda^*.$$

Hence $H_\Gamma \subset \Lambda^*/\Lambda = G_\Lambda$. The subgroup $H_\Gamma \subset G_\Lambda$ is isotropic, i.e., $q_\Lambda\,|H_\Gamma = 0$, since $\Gamma$ is an even lattice. Now we have

**Proposition 3.5.** *The correspondence $\Gamma \mapsto H_\Gamma$ is a one-to-one correspondence between even overlattices of $\Lambda$ and isotropic subgroups of $G_\Lambda$. Unimodular lattices correspond to isotropic subgroups $H$ with $|H|^2 = |G_\Lambda|$.*

*Proof.* The proof of the first assertion amounts to a simple verification, which can be left to the reader. The second assertion follows from the formula

$$\operatorname{disc}(\Gamma) = [\Gamma : \Lambda]^{-2} \operatorname{disc}(\Lambda)$$

(cf. Sect. 1.1). □

Two lattices $\Gamma, \Gamma' \subset \mathbb{R}^n$ are called *isomorphic* if there is an orthogonal automorphism $u \in \mathrm{O}_n(\mathbb{R})$ with $u(\Gamma) = \Gamma'$. An *automorphism* of the lattice $\Gamma$ is an orthogonal automorphism $u \in \mathrm{O}_n(\mathbb{R})$ with $u(\Gamma) = \Gamma$. Two overlattices $\Lambda \hookrightarrow \Gamma$ and $\Lambda \hookrightarrow \Gamma'$ are called *isomorphic* if there exists an automorphism of $\Lambda$ extending to an isomorphism of $\Gamma$ with $\Gamma'$.

In order to formulate the next result, we observe that an isomorphism $u : \Lambda_1 \to \Lambda_2$ of lattices induces an isomorphism $u^* : \Lambda_1^* \to \Lambda_2^*$ of the dual lattices and determines an isomorphism $\overline{u}: q_{\Lambda_1} \to q_{\Lambda_2}$ of their discriminant quadratic forms. In particular there is an induced homomorphism $\mathrm{O}(\Lambda) \to \mathrm{O}(q_\Lambda)$ between the automorphism groups of $\Lambda$ and $q_\Lambda$.

**Proposition 3.6.** *Two even overlattices $\Lambda \hookrightarrow \Gamma$ and $\Lambda \hookrightarrow \Gamma'$ are isomorphic if and only if the isotropic subgroups $H_\Gamma \subset G_\Lambda$ and $H_{\Gamma'} \subset G_\Lambda$ are conjugate under some automorphism of $\Lambda$.*

The proof of Proposition 3.6 is again a simple verification and can be left to the reader.

We now turn to the case that $\Lambda$ is a root lattice. We want to embed $\Lambda$ into an even unimodular lattice of the same dimension without creating new roots. That means that we are looking for even unimodular overlattices $\Lambda \hookrightarrow \Gamma$ such that $\Gamma \setminus \Lambda$ does not contain any root. This is translated into a condition on the isotropic subgroup $H_\Gamma$ as follows. We introduce a *length function* $l_\Lambda : G_\Lambda \to \mathbb{Q}$ on $G_\Lambda$ defined by

$$l_\Lambda(\xi) := \min\left\{x^2 \mid x \in \Lambda,\, \overline{x} = \xi\right\} \qquad \text{for } \xi \in G_\Lambda,$$

where $\overline{x}$ denotes the residue class of $x$ in $G_\Lambda$. Then $\Gamma \setminus \Lambda$ does not contain any root if and only if $l_\Lambda(\xi) \neq 2$ for all $\xi \in H_\Gamma$. Hence we get the following corollary of Propositions 3.5 and 3.6.

**Corollary 3.6.** *Let $\Lambda \subset \mathbb{R}^n$ be a root lattice. There is a natural one-to-one correspondence between isomorphism classes of even overlattices $\Lambda \hookrightarrow \Gamma$ with $\Gamma_2 = \Lambda_2$ and orbits of isotropic subgroups $H \subset G_\Lambda$ with $l_\Lambda(\xi) \neq 2$ for all $\xi \in H$ under the image of the natural homomorphism $\mathrm{O}(\Lambda) \to \mathrm{O}(q_\Lambda)$. Unimodular lattices correspond to subgroups $H$ with $|H|^2 = |G_\Lambda|$.*

**Example 3.1** Let $\Lambda = D_n$ for $n \equiv 0 \pmod 8$. Then we know from Sect. 1.4 that

$$\Lambda = \left\{(x_1,\ldots,x_n) \in \mathbb{Z}^n \;\middle|\; \sum x_i \text{ even}\right\},$$

$$G_\Lambda = \Lambda^*/\Lambda \cong (\mathbb{Z}/2\mathbb{Z}) \times (\mathbb{Z}/2\mathbb{Z}),$$

and $G_\Lambda$ is generated by $\omega_1$ and $\omega_2$, which are the classes of $(1,0,\ldots,0)$ and $v = \left(\frac{1}{2},\frac{1}{2},\ldots,\frac{1}{2}\right)$ respectively. Now

$$q_\Lambda(\omega_1) = 1, \qquad q_\Lambda(\omega_2) = 0, \qquad b_\Lambda(\omega_1,\omega_2) = \frac{1}{2},$$

$$l_\Lambda(\omega_1) = 1, \qquad l_\Lambda(\omega_2) = v^2 = \frac{n}{4}.$$

Thus for $n \geq 16$ there is a unique (up to conjugation under automorphisms of $\Lambda$) isotropic subgroup $H \subset G_\Lambda$ with $|H|^2 = |G_\Lambda|$ and $l_\Lambda(\xi) \neq 2$ for all $\xi \in H$, namely the subgroup $H = \{0,\omega_2\}$ of $G_\Lambda$. By Corollary 3.6 there is a unique (up to isomorphism) even unimodular lattice $\Gamma \subset \mathbb{R}^n$ with root sublattice $(\Gamma_2)_\mathbb{Z}$ equal to $D_n$, namely the lattice

$$D_n(v) := D_n \cup (D_n + v).$$

This shows in particular that for $n = 16$ there are up to isomorphism only two even unimodular lattices in $\mathbb{R}^n$, namely the lattices

$$E_8 \perp E_8 \qquad \text{and} \quad D_{16}(v).$$

This is a result due to Witt.

Analogously to the above theory for even lattices there is a theory for doubly even binary codes. The following results are due to H. Koch (cf. [46]).

Let $C \subset \mathbb{F}_2^n$ be a doubly even code of length $n$. The set

$$C_4 := \{c \in C \mid w(c) = 4\}$$

is called the *tetrad system* of $C$. A code $C$ generated by *tetrads*, i.e., by codewords $c \in C$ with $w(c) = 4$, is called a *tetrad code*. By Theorem 1.2 and Proposition 1.5, a tetrad code can be decomposed into irreducible tetrad codes, which up to equivalence are the codes $d_{2k}$, $k = 2, 3, \ldots$, which is the "double" of the even weight code of length $k$,

$$e_7 := H^{\perp}, \qquad e_8 := \widetilde{H},$$

where $H$ and $\widetilde{H}$ are the Hamming code and the extended Hamming code respectively.

Analogously to the case of lattices we define the finite abelian group

$$G_C := C^{\perp}/C,$$

and the weight function $w_C : G_C \to \mathbb{Z}$ by

$$w_C(\eta) := \min\{w(y) \mid \overline{y} = \eta\} \qquad \text{for } \eta \in G_C.$$

The symmetric group $\mathscr{S}_n$ operates by permutation of coordinates on $\mathbb{F}_2^n$. Recall that two codes $C$ and $C'$ in $\mathbb{F}_2^n$ are called *equivalent*, if there is a $\sigma \in \mathscr{S}_n$ with $\sigma(C) = C'$. The automorphism group $\operatorname{Aut}(C)$ of C is defined by

$$\operatorname{Aut}(C) := \{\sigma \in \mathscr{S}_n \mid \sigma(C) = C\}.$$

An automorphism $\sigma \in \operatorname{Aut}(C)$ induces an automorphism $\sigma^{\perp} \in \operatorname{Aut}(C^{\perp})$ of the dual code, and determines an automorphism $\overline{\sigma} \in \operatorname{Aut}(G_C)$ of $G_C$. Hence there is a canonical homomorphism $\operatorname{Aut}(C) \to \operatorname{Aut}(G_C)$.

**Proposition 3.7.** *Let $C$ be a tetrad code in $\mathbb{F}_2^n$. There is a natural one-to-one correspondence between equivalence classes of doubly even codes in $\mathbb{F}_2^n$ with tetrad system $C_4$ and orbits of subgroups $H$ of $G_C$ with $w_C(\xi) \in 4\mathbb{Z} - \{4\}$ for all $\xi \in H$ under the image of the natural homomorphism* $\operatorname{Aut}(C) \to \operatorname{Aut}(G_C)$. *Self-dual codes correspond to subgroups $H$ with $|H|^2 = |G_C|$.*

The proof of Proposition 3.7 is completely analogous to the proof of Corollary 3.6.

In Sect. 1.3 we associated to a code $C \subset \mathbb{F}_2^n$ a lattice $\Gamma_C \subset \mathbb{R}^n$. Recall that $\Gamma_C = \frac{1}{\sqrt{2}}\rho^{-1}(C)$, where $\rho : \mathbb{Z}^n \to \mathbb{F}_2^n$ is the reduction mod 2. Let $(\varepsilon_1, \ldots, \varepsilon_n)$ be the standard basis of $\mathbb{R}^n$. Then $\frac{1}{\sqrt{2}}2\varepsilon_i \in \Gamma_C$ for all $1 \leq i \leq n$. These are $n$ pairwise orthogonal roots. Thus $\Gamma_C$ contains a root lattice of type $nA_1$. The results of Sect. 1.3, Corollary 3.6 and Proposition 3.7 can be combined to yield the following theorem.

**Theorem 3.3.** *The correspondence $C \mapsto \Gamma_C$ induces a one-to-one correspondence between equivalence classes of doubly even codes $C$ in $\mathbb{F}_2^n$ and isomorphism classes of even lattices in $\mathbb{R}^n$ containing a root lattice of type $nA_1$. Self-dual codes correspond to unimodular lattices.*

*Proof.* This follows from Corollary 3.6 and Proposition 3.7. The details of the proof can be left as an exercise to the reader. □

## 3.4 The Classification of Even Unimodular Lattices of Dimension 24

In Sect. 3.2 we classified the possible root systems in an even unimodular lattice of dimension 24. In the last section we saw that the problem of classifying the even unimodular lattices containing a given root lattice of the same dimension as the root lattice can be translated into the problem of classifying certain subgroups of finite abelian groups. Theorem 3.3 tells us that even unimodular lattices in $\mathbb{R}^n$ containing the root lattice $nA_1$ correspond to doubly even self-dual codes in $\mathbb{F}_2^n$.

**Example 3.2** One of the root lattices of Proposition 3.4 is the root lattice $24A_1$. Let $\Gamma \subset \mathbb{R}^{24}$ be an even unimodular lattice containing $24A_1$ as the root sublattice, i.e., containing no other roots than the roots of $24A_1$. Then by Theorem 3.3 $\Gamma = \Gamma_C$ for a doubly even self-dual code $C \subset \mathbb{F}_2^{24}$ without tetrads. We showed the existence of such a code in Sect. 2.8. It is the extended Golay code $\widetilde{G} \subset \mathbb{F}_2^{24}$. In Sect. 2.8 we also showed that such a code is unique up to equivalence. Hence by Theorem 3.3 such a lattice $\Gamma$ exists and is unique up to isomorphism.

Now, for each of the root lattices $\Lambda$ of Proposition 3.4, it is possible to show the existence and uniqueness of the corresponding subgroup of $G_\Lambda$, respectively of the corresponding code. Thus one can prove:

**Theorem 3.4 (Niemeier).** *Up to isomorphism there exist precisely 24 even unimodular lattices in $\mathbb{R}^{24}$. Each lattice is uniquely determined by its root sublattice. The possible root sublattices are the 24 listed in Proposition 3.4.*

This result was obtained by H.-V. Niemeier in 1973 [68]. Niemeier used among other things a method of M. Kneser [44], which involves studying neighbours of lattices (see Sect. 4.1), requires extensive calculations, and does not explain the appearance of the strange list of root lattices. This list was explained by B. B. Venkov [89]. The outline of the proof of Theorem 3.4 above is due to Venkov [89].

Using Proposition 1.5, we obtain the following corollary of Theorem 3.3 and Theorem 3.4.

**Corollary 3.7.** *From the (up to isomorphism) 24 even unimodular lattices in $\mathbb{R}^{24}$, 9 correspond to doubly even self-dual codes in $\mathbb{F}_2^{24}$. They correspond to the root lattices*

$$24A_1,\ 6D_4,\ 4D_6,\ 3D_8,\ 2D_{12},\ D_{24},\ 3E_8,\ 2E_7 \perp D_{10},\ E_8 \perp D_{16}.$$

Some other lattices are obtained by considering codes over $\mathbb{F}_p$, see Sect. 5.4.

The 9 lattices of Corollary 3.7 correspond to the doubly even self-dual codes in $\mathbb{F}_2^{24}$ with tetrad systems

$$\emptyset,\ 6d_4,\ 4d_6,\ 3d_8,\ 2d_{12},\ d_{24},\ 3e_8,\ 2e_7+d_{10},\ e_8+d_{16}.$$

By Theorem 3.3, the fact that each of the corresponding even unimodular lattices in $\mathbb{R}^{24}$ is uniquely determined by its root sublattice follows from the fact that the corresponding code is characterized by its tetrad system. H. Koch has shown that the above 9 codes satisfy a certain completeness principle and he has used this to show that each of these codes is uniquely determined by its tetrad system, see [47] and the following exercise.

**Exercise 3.4** Let us consider the case $6D_4$. The corresponding lattice is given by a doubly even self-dual code $C \subset \mathbb{F}_2^{24}$ with tetrad system $C_4$ of type $6d_4$. We denote by $\langle C_4\rangle$ the code generated by $C_4$, hence $\langle C_4\rangle = 6d_4$. We have

$$C/\langle C_4\rangle \subset G_{\langle C_4\rangle} = \langle C_4\rangle^{\perp}/\langle C_4\rangle \cong \mathbb{F}_4^6$$

where $\mathbb{F}_4 = \{0,1,\omega,\overline{\omega}\}$ with $1 = \overline{(1,1,0,0)}$, $\omega = \overline{(1,0,1,0)}$, $\overline{\omega} = \overline{(1,0,0,1)}$. Here $\overline{x}$ denotes the equivalence class of an element $x \in \mathbb{F}_2^4$ in $d_4^{\perp}/d_4$. Show that $\langle C_4\rangle^{\perp}/\langle C_4\rangle$ is the hexacode in $\mathbb{F}_4^6$ (cf. Sect. 2.8). (Hint: By Proposition 2.7 there are 735 codewords of weight 8 in $C$, which are either represented in the form $(0,0,a_1,a_2,a_3,a_4)$, $a_i \in \mathbb{F}_4^*$, and permutations of the positions in $\mathbb{F}_4^6$ or belong to 15 codewords of weight 8 in $\langle C_4\rangle$. There can be at most 3 words of the form $(0,0,a_1,a_2,a_3,a_4)$, $a_i \in \mathbb{F}_4^*$, in $C/\langle C_4\rangle$. For each codeword of this form yields 16 codewords of weight 8 in $C$. There are 15 possibilities for the distribution of zeros. But there can be no more than $16\cdot 3\cdot 15 = 720$ codewords of weight 8 in $C/\langle C_4\rangle$. Therefore there are exactly 3 codewords of the form $(0,0,a_1,a_2,a_3,a_4)$, $a_i \in \mathbb{F}_4^*$, in $C/\langle C_4\rangle$. After a possible permutation of the positions in $\mathbb{F}_2^{24}$, we may assume that these are the words $(0,0,1,1,1,1)$, $(0,0,\omega,\omega,\omega,\omega)$, and $(0,0,\overline{\omega},\overline{\omega},\overline{\omega},\overline{\omega})$. In particular, the last row of the generator matrix of the hexacode in Sect. 2.8 is in $C/\langle C_4\rangle$. First show that also the 6 words $(1,1,1,1,0,0)$, $(\omega,\omega,\omega,\omega,0,0)$, $(\overline{\omega},\overline{\omega},\overline{\omega},\overline{\omega},0,0)$, $(1,1,0,0,1,1)$, $\omega,\omega,0,0,\omega,\omega)$, and $(\overline{\omega},\overline{\omega},0,0,\overline{\omega},\overline{\omega})$ are in $C/\langle C_4\rangle$, possibly after another permutation of the positions in $\mathbb{F}_2^{24}$. Then show that also the first and second row of the generator matrix of the hexacode are in $C/\langle C_4\rangle$.)

Note that we obtained in particular in Sect. 3.2 and Sect. 3.3 the classification of Witt in dimension 16: There are (up to isomorphism) exactly two even unimodular lattices in $\mathbb{R}^{16}$. Each lattice is uniquely determined by its root sublattice. The possible root sublattices are $E_8+E_8$ and $D_{16}$. Both even unimodular lattices correspond to doubly even self-dual codes of length 16.

So we have classified even unimodular lattices up to dimension 24. One can derive from the Minkowski-Siegel mass formula that the number of inequivalent even unimodular lattices of dimension 32 is greater than $80{,}000{,}000$ [81, p. 55] (but see also [48]). Doubly even self-dual codes are classified up to length 32 [16]. The number of inequivalent doubly even self-dual binary codes of length 40 is greater than $17{,}000$.

# Chapter 4
# The Leech Lattice

## 4.1 The Uniqueness of the Leech Lattice

In Sect. 2.8 we defined an even unimodular lattice $\Lambda_{24} \subset \mathbb{R}^{24}$ with $x \cdot x \geq 4$ for all $0 \neq x \in \Lambda_{24}$: the Leech lattice. This chapter is devoted to this important lattice. We shall first show the uniqueness of this lattice. We shall prove the following theorem.

**Theorem 4.1 (Conway).** *The Leech lattice is the unique (up to isomorphism) even unimodular lattice in $\mathbb{R}^{24}$ without roots.*

This characterization of the Leech lattice was given by J. H. Conway [11] in 1969. The following proof, which uses the classification of even unimodular lattices in $\mathbb{R}^{24}$ with nontrivial root systems (see Chapter 3), was given by Venkov [89] (= [21, Chap. 18]).

The proof will involve the concept of a neighbour of a lattice (cf. [44]).

**Definition.** Two lattices $\Lambda$ and $\Gamma$ in $\mathbb{R}^n$ are called *neighbours* if their intersection $\Lambda \cap \Gamma$ has index 2 in each of them.

Let $\Lambda \subset \mathbb{R}^n$ be an even unimodular lattice. Then a neighbour of $\Lambda$ can be constructed as follows. Take any vector $u \in \Lambda$ satisfying

$$\frac{u^2}{4} \in \mathbb{Z}, \qquad \frac{u}{2} \notin \Lambda,$$

and define

$$\begin{aligned} \Lambda_1 &:= \{x \in \Lambda \mid x \cdot u \equiv 0 \ (\mathrm{mod}\ 2)\}, \\ \Lambda^u &:= \Lambda_1 \cup \left(\frac{u}{2} + \Lambda_1\right). \end{aligned}$$

Then it is not difficult to show that $\Lambda_1 = \Lambda \cap \Lambda^u$, $\Lambda^u$ is unimodular, $\Lambda$ and $\Lambda^u$ are neighbours, all neighbours of $\Lambda$ arise in this way, and $u$, $u'$ produce the same neighbour if and only if $\frac{u}{2} \equiv \frac{u'}{2} \ (\mathrm{mod}\ \Lambda_1)$. Note that the construction of the Leech lattice in Sect. 2.8 was a special case of this construction.

*Proof of Theorem 4.1.* Suppose that $\Lambda \subset \mathbb{R}^{24}$ is an even unimodular lattice with $\Lambda_2 = \emptyset$. We want to prove that $\Lambda$ is isomorphic to the Leech lattice. The theta function of $\Lambda$ is a modular form of weight 12. By Corollary 2.2, the condition $\Lambda_2 = \emptyset$ implies that the theta function of $\Lambda$ is the same as the theta function of the Leech lattice. For the space $M_{12}$ of modular forms of weight 12 is 2-dimensional, and we have two conditions for the theta function of $\Lambda$: $a_0 = |\Lambda_0| = 1$, $a_1 = |\Lambda_2| = 0$, where

$$\vartheta_\Lambda(\tau) = \sum_{r=0}^{\infty} a_r q^r, \qquad q = e^{2\pi i \tau},$$

(cf. also Lemma 2.5). This implies in particular that $\Lambda_8 \neq \emptyset$.

We now consider a neighbour of the lattice $\Lambda$. Let $u \in \Lambda_8$, $\frac{u}{2} \notin \Lambda$. Consider the neighbouring lattice $\Gamma := \Lambda^u = \Lambda_1 \cup \left(\frac{u}{2} + \Lambda_1\right)$ where $\Lambda_1 = \{x \in \Lambda \mid x \cdot u \equiv 0 \pmod 2\}$. By the remarks above, $\Gamma$ is an even unimodular lattice in $\mathbb{R}^{24}$. The set $\Gamma_2$ of roots in $\Gamma$ is non-empty, since $\frac{u}{2} \in \Gamma_2$. We shall prove that the root sublattice of $\Gamma$ is equal to $kA_1$ for some $k$. For this purpose, it suffices to show that any two roots $x, y \in \Gamma_2$ with $x \neq \pm y$ are orthogonal to each other. Let $x, y$ be two roots of $\Gamma$ with $x \neq \pm y$. Assume that $x \cdot y \neq 0$. By the Cauchy-Schwarz inequality, $x \cdot y = \pm 1$ (cf. the remarks at the beginning of Sect. 1.4). Changing the sign of $y$ if necessary, we may assume that $x \cdot y = 1$. This implies that $x - y$ is a root. But since $\Lambda_2 = \emptyset$, we have $x, y \in \frac{u}{2} + \Lambda_1$, and hence $x - y \in \Lambda_1 \subset \Lambda$, which is a contradiction. This shows that $(\Gamma_2)_{\mathbb{Z}} = kA_1$ for some $k$. Then, according to Proposition 3.4, we must have $k = 24$. By Example 3.2, $\Gamma$ is isomorphic to $\Gamma_{\widetilde{G}}$ for the extended Golay code $\widetilde{G}$. Therefore we can choose elements $e_1, \ldots, e_{24}$ in $\mathbb{R}^{24}$ such that $e_i \cdot e_j = 0$ if $i \neq j$, $e_i^2 = \frac{1}{2}$, and $e_1, \ldots, e_{24}$ is a basis of $\mathbb{R}^{24}$. Moreover, $\sum n_i e_i \in \Gamma$ if and only if $n_i \in \mathbb{Z}$ for all $1 \leq i \leq 24$ and $\rho\left((n_1, n_2, \ldots, n_{24})\right) \in \widetilde{G}$.

Since $\Lambda$ and $\Gamma$ are neighbours and any neighbour of $\Gamma$ is of the form $\Gamma^v$ for a suitable $v \in \Gamma$, there exists a $v \in \Gamma$ with $\frac{v^2}{4} \in \mathbb{Z}$ and $\frac{v}{2} \notin \Gamma$ such that $\Lambda = \Gamma^v = \Gamma_1 \cup (\frac{v}{2} + \Gamma_1)$ where $\Gamma_1 = \{x \in \Gamma \mid x \cdot v \equiv 0 \pmod 2\}$. Then $v = \sum m_i e_i$ with $m_i \in \mathbb{Z}$ for $i = 1, \ldots, 24$. Each $m_i$ must be odd. For assume that $m_j$ is even. Then $(2e_j) \cdot v = m_j \equiv 0 \pmod 2$ and hence $2e_j \in \Gamma_1 \subset \Gamma^v = \Lambda$. But $(2e_j)^2 = 2$ and hence $2e_j \in \Lambda_2$, which is absurd, since $\Lambda_2 = \emptyset$. Therefore each $m_i$ is odd. By the remarks on neighbouring lattices, $v$ is defined modulo $2\Gamma_1$. Hence we may assume that $m_i \equiv \pm 1 \pmod 4$. Since $s_{2e_i}(e_j) = (1 - 2\delta_{ij})e_j$ for $1 \leq i, j \leq 24$, by applying appropriate automorphisms $s_{2e_i}$ we can even arrange that $m_i \equiv 1 \pmod 4$. For all $1 \leq i, j \leq 24$ we have $4e_i + 4e_j \in 2\Gamma_1$, in particular also $8e_i \in 2\Gamma_1$. Therefore the $m_i$ can be chosen (modulo $2\Gamma_1$) in such a way that $m_1 \in \{1, 5\}$ and $m_i = 1$ for $2 \leq i \leq 24$. For $v = e_1 + \ldots + e_{24}$ we would have $(\frac{v}{2})^2 = 3$, which is not possible, since $\frac{v}{2} \in \Lambda$ and $\Lambda$ is an even lattice. Therefore we must have $m_1 = 5$ and hence

$$v = 5e_1 + e_2 + \ldots e_{24} = 4e_1 + \frac{1}{\sqrt{2}} \cdot \mathbf{1}.$$

But in Sect. 2.8 we have constructed the Leech lattice as the lattice $\Gamma^v$ for such a $v$. Hence $\Gamma^v$ is the Leech lattice. This completes the proof of Theorem 4.1. □

**Remark 4.1** There is a minor error in the original proof of Venkov which was also copied in the first edition of this book, see [91].

## 4.2 The Sphere Covering Determined by the Leech Lattice

In this section we shall consider the sphere packing and covering determined by the Leech lattice.

Let $\Gamma \subset \mathbb{R}^n$ be a lattice. The minimum squared distance of any two lattice points is

$$d_\Gamma = \min_{\substack{x\in\Gamma \\ x\neq 0}} x\cdot x\,.$$

This is called the *minimum squared distance* of the lattice. If we draw spheres of radius

$$r = \frac{1}{2}\sqrt{d_\Gamma}$$

around the points of $\Gamma$ we obtain a *sphere packing* in $\mathbb{R}^n$, namely infinitely many non-overlapping $(n-1)$-spheres in $\mathbb{R}^n$. The number $r$ is called the *packing radius* of $\Gamma$.

The packing radius of the Leech lattice $\Lambda_{24}$ is equal to 1, and the corresponding sphere packing is very dense. It is the densest known sphere packing in $\mathbb{R}^{24}$. The Leech lattice was discovered by J. Leech in connection with the sphere packing problem [54].

The sphere packing problem is the problem to find the densest sphere packing. Closely related to the packing problem is the so called *kissing number* or *Newton number problem*: Given an $(n-1)$-dimensional sphere, what is the maximal number $\tau_n$ of (non-overlapping) $(n-1)$-dimensional spheres of the same radius touching it? It is easy to show that $\tau_1 = 2$, $\tau_2 = 6$, but already difficult to show that $\tau_3 = 12$. (There was a famous controversy about this number between Isaac Newton and David Gregory.) We have shown that

$$\tau_8 \geq 240\,,$$

since there are 240 roots in the lattice $E_8$, and

$$\tau_{24} \geq 196560\,,$$

since there are 196560 vectors of squared length 4 in $\Lambda_{24}$ (cf. Sect. 2.7). A. M. Odlyzko and N. J. A. Sloane ([73], cf. also [21, Chapter 13]), have shown that the above inequalities are in fact equalities. These are all the known Newton numbers.

Dual to the packing problem is the *covering problem*: Find the least dense covering of $\mathbb{R}^n$ by overlapping spheres! The packing radius $r$ of a lattice $\Gamma \subset \mathbb{R}^n$ was defined above. It is the largest number $\rho$, such that spheres of radius $\rho$ centred at the lattice points do not overlap. The *covering radius* $R$ of a lattice $\Gamma \subset \mathbb{R}^n$ is the smallest number $\rho$, such that spheres of radius $\rho$ centred at the lattice points of $\Gamma$ cover the whole $\mathbb{R}^n$. In this section we want to give an outline of the determination of the covering radius of the Leech lattice due to J. H. Conway, R. A. Parker, and N. J. A. Sloane ([17], see also [21, Chap. 23]). For this purpose one investigates the holes of the Leech lattice.

Let $\Gamma \subset \mathbb{R}^n$ be a lattice. A point $c \in \mathbb{R}^n$ whose distance from the points of $\Gamma$ is a local maximum is called a *hole* of $\Gamma$. Let $c$ be a hole of $\Gamma$, and let $d$ be the minimum distance of $c$ from the points of $\Gamma$. Then there are at least $n+1$ points of $\Gamma$ having distance $d$ from $c$, and these lattice points form the vertices of a convex polytope $P$ of dimension $n$ which is

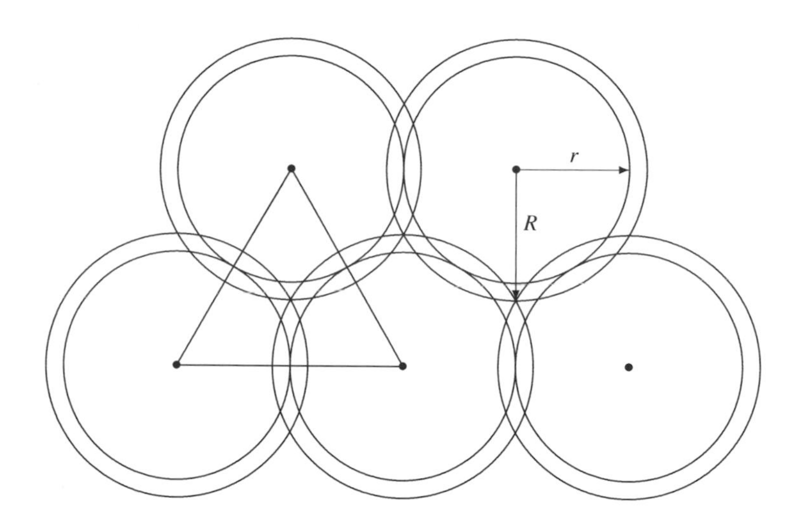


**Fig. 4.1** Packing radius $r$, covering radius $R$, and Delaunay polytope

called the *Delaunay polytope* of $c$. Two holes are *equivalent*, if their Delaunay polytopes are congruent under translations and automorphisms of the lattice $\Gamma$.

A hole is called *deep* if $d$ is equal to the covering radius $R$ of $\Gamma$. Otherwise it is called *shallow*.

We want to determine the deep holes of the Leech lattice $\Lambda_{24}$. We first need two lemmas. The first one can be found in [12] (see also [21, Chap. 10]). For the second one see [17]([21, Chap. 23]).

**Lemma 4.1 (Conway).** *Every element of $\Lambda_{24}$ is congruent modulo $2\Lambda_{24}$ to an element of squared length at most 8.*

*Proof.* We count the number of congruence classes modulo $2\Lambda_{24}$ of vectors of squared length at most 8. Let $a_r$ denote the number of elements $x \in \Lambda_{24}$ with $x^2 = 2r$. From the theta function of $\Lambda_{24}$, which is by Sect. 2.7

$$\vartheta_{\Lambda_{24}} = E_4^3 - 720\Delta\,,$$

we get

$$\begin{aligned} a_1 &= 0, \\ a_2 &= 196560, \\ a_3 &= 16773120, \\ a_4 &= 398034000. \end{aligned}$$

So there are $a_2 + a_3 + a_4$ vectors of squared length at most 8.

Now let $x,y \in \Lambda_{24}$ be two vectors with $x^2, y^2 \leq 8$, $x \neq \pm y$, which are congruent modulo $2\Lambda_{24}$. Then since $x \pm y \in 2\Lambda_{24}$ and $\Lambda_{24}$ contains no roots we must have $(x \pm y)^2 = x^2 \pm 2xy + y^2 \geq 16$. But since $x^2$ and $y^2$ are both at most 8, this is only possible if $x \cdot y = 0$ and $x^2 = y^2 = 8$. In particular if $x^2 \leq 6$ then $x$ is only congruent to $-x$ modulo $2\Lambda_{24}$. If $x^2 = 8$ then $x$ can only be congruent modulo $2\Lambda_{24}$ to $-x$ and to vectors orthogonal to $x$. If $y$ and $z$ are such vectors then $y$ and $z$ have to be orthogonal to each other, too. Therefore the vectors congruent to $x$ modulo $2\Lambda_{24}$ but different from $\pm x$ must be orthogonal to $x$ and mutually orthogonal. There are at most 46 such elements. Therefore a congruence class of some vector $x$ with $x^2 = 8$ contains at most 48 elements. We have therefore at least

$$\frac{a_2}{2} + \frac{a_3}{2} + \frac{a_4}{48}$$

distinct classes of $\Lambda_{24}/2\Lambda_{24}$. But this sum turns out to be equal to $2^{24}$ which is the order of $\Lambda_{24}/2\Lambda_{24}$. Hence every element of $\Lambda_{24}/2\Lambda_{24}$ is represented by an element of squared length at most 8. This proves Lemma 4.1. □

**Lemma 4.2.** *Let $c$ be a deep hole of $\Lambda_{24}$ with Delaunay polytope $P$. Let $v_1, \ldots, v_\nu$ be the vertices of $P$. Then $(v_i - v_j)^2 = 4, 6,$ or $8$, for $i \neq j$.*

*Proof.* Since the $v_i$ are in $\Lambda_{24}$, we have $(v_i - v_j)^2 \geq 4$ for all $i \neq j$. Now assume that $i \neq j$ and $(v_i - v_j)^2 > 8$. By Lemma 4.1 there exist $x, y \in \Lambda_{24}$ with $x^2 \leq 8$ and $v_i - v_j = x + 2y$. Set $x' := v_i - y$, $x'' := v_j + y$, and $z := \frac{1}{2}(v_i + v_j)$. Then $x = x' - x''$ and $z = \frac{1}{2}(x' + x'')$. Since $v_i$ and $v_j$ have the same distance from $c$, we get

$$(v_i - c)^2 = (v_i - z)^2 + (z - c)^2 = \frac{1}{4}(v_i - v_j)^2 + (z - c)^2.$$

By interchanging $x'$ and $x''$ if necessary, we may assume that $(x' - x'') \cdot (z - c) \leq 0$. Since $(v_i - v_j)^2 > 8$ by assumption, we have

$$\begin{aligned}
(x' - c)^2 &= (x' - z)^2 + 2(x' - z) \cdot (z - c) + (z - c)^2 \\
&= \frac{1}{4}x^2 + (x' - x'') \cdot (z - c) + (z - c)^2 \\
&< \frac{1}{4}(v_i - v_j)^2 + (z - c)^2 \\
&= (v_i - c)^2.
\end{aligned}$$

But this means that $x'$ is closer to $c$ than $v_i$, which is impossible. Thus $(v_i - v_j)^2 = 4, 6,$ or $8$, for $i \neq j$. This proves Lemma 4.2. □

Let $c$ be a deep hole of $\Lambda_{24}$, and $v_1, \ldots, v_\nu$ be the vertices of its Delaunay polytope $P$. By Lemma 4.2, $(v_i - v_j)^2 = 4, 6,$ or $8$, for $i \neq j$. Conway, Parker, and Sloane associate a graph to $c$ as follows. The vertices are in one-to-one correspondence with the vectors $v_i$ (and are denoted by the same symbols). Two vertices $v_i$ and $v_j$ are

(i) not joined if $(v_i - v_j)^2 = 4$,
(ii) joined by an edge if $(v_i - v_j)^2 = 6$,

(iii) joined by two edges if $(v_i - v_j)^2 = 8$.

The resulting graph is called the *hole diagram* corresponding to $c$.

Now suppose that $R = \sqrt{2}$. Then $(v_i - c)^2 = 2$ for all $1 \le i \le \nu$, and

(i) $(v_i - v_j)^2 = 4 \Longleftrightarrow (v_i - c) \cdot (v_j - c) = 0$
(ii) $(v_i - v_j)^2 = 6 \Longleftrightarrow (v_i - c) \cdot (v_j - c) = -1$
(iii) $(v_i - v_j)^2 = 8 \Longleftrightarrow (v_i - c) \cdot (v_j - c) = -2$

So in this case the hole diagram is the Coxeter-Dynkin diagram corresponding to the set of vectors $\{v_i - c \mid 1 \le i \le \nu\}$.

The notion of an extended Coxeter-Dynkin diagram was introduced in Sect. 1.5. We have the following two propositions (cf. [17, Theorem 5] and [17, Theorem 6]).

**Proposition 4.1.** *A hole diagram with no extended Coxeter-Dynkin diagram embedded in it contains only ordinary Coxeter-Dynkin diagrams as components.*

*Proof.* This follows from the proof of Theorem 1.2, i.e., from the classification of irreducible root lattices. □

**Proposition 4.2.** *The radius of a hole with a diagram in which all components are ordinary Coxeter-Dynkin diagrams is less than $\sqrt{2}$.*

*Proof.* Such a diagram corresponds to a fundamental system $(e_1, \dots, e_\nu)$ of roots of a root lattice $\Gamma$ in $\mathbb{R}^\nu$. By the definition of the hole diagram, the vectors $e_1, \dots, e_\nu$ have the same mutual distances as the points $v_1, \dots, v_\nu$. Therefore we can identify the points $v_1, \dots, v_\nu$ with the endpoints of the vectors $e_1, \dots, e_\nu$. Let $H$ be the affine hyperplane of $\mathbb{R}^\nu$ containing these points. In $H$ there is also a point corresponding to the centre $c$ of the hole. But the origin in $\mathbb{R}^\nu$ does not lie in $H$, since it is not a linear combination of $e_1, \dots, e_\nu$. The origin has distance $\sqrt{2}$ from the endpoints of the vectors $e_1, \dots, e_\nu$. Thus $|v_i - c| = \sqrt[2]{(v_i - c)^2} < \sqrt{2}$ for each $i \in \{1, \dots, \nu\}$ (cf. Fig. 4.2). This proves Proposition 4.2. □

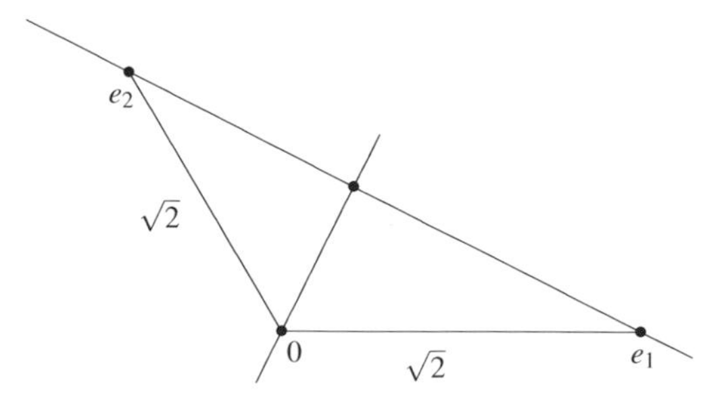


**Fig. 4.2** See the proof of Proposition 4.2

Now Conway, Parker, and Sloane proved the following two theorems ([17], see also [21, Chap. 23]).

**Theorem 4.2 (Conway, Parker, Sloane).** *The covering radius of the Leech lattice is* $R = \sqrt{2}$.

**Theorem 4.3 (Conway, Parker, Sloane).** *There are 23 inequivalent deep holes in the Leech lattice under congruences of that lattice, and they are in one-to-one correspondence with the 23 even unimodular lattices in* $\mathbb{R}^{24}$ *containing roots. The components of the hole diagram of a deep hole are the extended Coxeter-Dynkin diagrams of the irreducible components of the root sublattice of the corresponding even unimodular lattice.*

Conway, Parker, and Sloane derived these theorems from the preceding results by a rather long case-by-case consideration. Later R. E. Borcherds gave a uniform proof of these results [5]. We indicate Borcherds' proof in Sect. 4.4.

## 4.3 Twenty-Three Constructions of the Leech Lattice

Let $\Gamma \subset \mathbb{R}^n$ be an even overlattice of a direct sum of lattices

$$\Gamma_1 \oplus \Gamma_2 \oplus \ldots \oplus \Gamma_k .$$

By Sect. 3.3 such a lattice is obtained as follows. The lattice $\Gamma$ is generated by $\Gamma_1 \oplus \Gamma_2 \oplus \ldots \oplus \Gamma_k$ together with certain vectors

$$y = y_1 + y_2 + \ldots + y_k ,$$

where each $y_i$ lies in $\Gamma_i^*$. These vectors $y$ are called *glue vectors*, and we say that the lattice $\Gamma$ is obtained by *gluing* the components $\Gamma_1, \ldots, \Gamma_k$ together by the glue vectors $y$.

In particular the 23 even unimodular lattices in $\mathbb{R}^{24}$ containing roots can be obtained by gluing certain irreducible root lattices together by certain glue vectors. For abbreviation let us call these 23 lattices described by Niemeier the *Niemeier lattices*.

Now Conway and Sloane ([20], see also [21, Chapter 24]) gave a construction of the Leech lattice from each Niemeier lattice, using gluing theory. In order to state these results, we need some preparations. We follow [5].

Let $\Gamma \subset \mathbb{R}^n$ be an irreducible root lattice, and let $(e_1, \ldots, e_n)$ be a fundamental system of roots of $\Gamma$. Recall that the Weyl vector of $\Gamma$ is the vector

$$\rho = \frac{1}{2} \sum_{\alpha > 0} \alpha,$$

where the sum is over the positive roots of $\Gamma$ (with respect to the given fundamental system of roots of $\Gamma$), cf. Sect. 1.5.

Now let $N$ be a Niemeier lattice and choose a fundamental system of roots for its root sublattice. The *Weyl vector* $\rho$ of $N$ is half the sum of the positive roots of $N$ (with respect to the given fundamental system of roots). It is the sum of the Weyl vectors of the irreducible components of the root sublattice of $N$. By Corollary 3.4 the root sublattice of

$N$ has the same rank as $N$, and by Corollary 3.5 all its irreducible components have the same Coxeter number $h$. We call $h$ the *Coxeter number* of the Niemeier lattice $N$.

**Lemma 4.3.** *The squared length of the Weyl vector $\rho$ of $N$ is $2h(h+1)$.*

*Proof.* Let $n_1, \ldots, n_k$ be the ranks of the irreducible components of the root sublattice of $N$, and let $h$ be the Coxeter number of such a component. By Lemma 1.17 we get

$$\rho^2 = \sum_{i=1}^{k} \frac{1}{12} n_i h(h+1) = 2h(h+1),$$

since $n_1 + \ldots + n_k = 24$. This proves Lemma 4.3. □

For the following two lemmas see also [5, Lemma 2.6] and [5, Lemma 2.7].

**Lemma 4.4.** *The Weyl vector $\rho$ of $N$ lies in $N$.*

*Proof.* Since $N$ is unimodular, it suffices to show that $\rho$ lies in the dual lattice $N^*$. This means that we have to show that $\rho \cdot y$ is an integer for all $y \in N$. By definition

$$\rho = \frac{1}{2} \sum_{\alpha > 0} \alpha,$$

where the sum is over all positive roots $\alpha$ of $N$. Therefore

$$\begin{aligned}
(2\rho \cdot y)^2 &= \left( \left( \sum_{\alpha>0} \alpha \right) \cdot y \right)^2 \\
&= \left( \sum_{\alpha>0} \alpha \cdot y \right)^2 \\
&\equiv \sum_{\alpha>0} (\alpha \cdot y)^2 \pmod 2.
\end{aligned}$$

By Proposition 1.6

$$\sum_{\alpha>0} (\alpha \cdot y)^2 = y^2 h \equiv 0 \pmod 2,$$

as $y^2$ is even. Therefore the term $(2\rho \cdot y)^2$ is an even integer. Since $\rho \cdot y$ is rational, $\rho \cdot y$ has to be an integer, which proves Lemma 4.4. □

**Lemma 4.5.** *For $y \in N$ we have*

$$\left( \frac{\rho}{h} - y \right)^2 \geq 2 \left( 1 + \frac{1}{h} \right).$$

*The vectors $y \in N$ for which equality holds form a complete set of representatives for $N/\Gamma$, where $\Gamma$ is the root sublattice of $N$.*

*Proof.* Since $\rho^2 = 2h(h+1)$, we have

$$\begin{aligned}
\left(\frac{\rho}{h} - y\right)^2 - 2\left(1 + \frac{1}{h}\right) &= \frac{(\rho - hy)^2 - \rho^2}{h^2} \\
&= \frac{hy^2 - 2\rho \cdot y}{h} \\
&= \frac{\sum_{\alpha>0} (y \cdot \alpha)^2 - \sum_{\alpha>0} y \cdot \alpha}{h} \quad \text{(by Proposition 1.6)} \\
&= \sum_{\alpha>0} \frac{(y \cdot \alpha)^2 - y \cdot \alpha}{h},
\end{aligned}$$

where the sums are over all positive roots $\alpha$ of $N$. The last sum is greater than or equal to 0 because the numbers $y \cdot \alpha$ are integers. This proves the first part of the lemma.

The above sum is equal to zero if and only if $y \cdot \alpha$ is 0 or 1 for all positive roots $\alpha$ of $N$. Let

$$\Gamma = \Gamma_1 \perp \Gamma_2 \perp \ldots \perp \Gamma_k$$

be the decomposition of the root sublattice $\Gamma$ into irreducible components. Then a vector $y \in N$ can be written

$$y = y_1 + y_2 + \ldots + y_k,$$

with certain vectors $y_i \in \Gamma_i^*$. If $y \cdot \alpha$ is 0 or 1 for all positive roots $\alpha$ of $N$, then each $y_i \in \Gamma_i^*$ has inner product 0 or 1 with all positive roots of $\Gamma_i$. By Lemma 1.18 the vectors $y_i \in \Gamma_i^*$ with this property form a complete set of representatives for $\Gamma_i^*/\Gamma_i$. Therefore the vectors $y \in N$ that have inner product 0 or 1 with all positive roots of $N$ form a complete set of representatives for $N/\Gamma$, and this proves the last part of Lemma 4.5. □

We are now ready to indicate the construction of Conway and Sloane. Let $N$ be a Niemeier lattice with root sublattice $\Gamma$. Choose a fundamental system of roots of $\Gamma$ whose Weyl vector is $\rho$. Let $\{f_1, \ldots, f_\nu\}$ be the union of the fundamental system of roots and the set of the corresponding highest roots for the irreducible components of $\Gamma$. Let $\{g_1, \ldots, g_\mu\}$ be the set of vectors of the form $g_i = y_i - \frac{\rho}{h}$, where $y_i$ is any vector of $N$ such that $g_i$ has squared length $2(1 + \frac{1}{h})$. By Lemma 4.5, the vectors $y_i$ form a complete set of coset representatives for $N/\Gamma$. In particular we have $\mu = \sqrt{\text{disc}\,(\Gamma)}$. The set $\{g_1, \ldots, g_\mu\}$ will be the set of glue vectors.

Now Conway and Sloane proved the following theorem:

**Theorem 4.4 (Conway, Sloane).** *With the above notations one has:*
(i) *The vectors $\sum m_i f_i + \sum n_j g_j$ with $\sum n_j = 0$ form the lattice $N$.*
(ii) *The vectors $\sum m_i f_i + \sum n_j g_j$ with $\sum m_i + \sum n_j = 0$ form a copy of the Leech lattice.*

Part (i) follows because $\{f_1, \ldots, f_\nu\}$ is a system of generators for $\Gamma$ and $\{y_1, \ldots, y_\mu\}$ is a complete set of coset representatives for $N/\Gamma$. For a proof of (ii) we refer to [5].

In particular for the Niemeier lattice with root sublattice of type $24A_1$ we get the construction of the Leech lattice of Sect. 2.8.

## 4.4 Embedding the Leech Lattice in a Hyperbolic Lattice

In this section we indicate Borcherds' proof of Theorem 4.2 and Theorem 4.3 [5].

Up to now, we have only considered lattices in $\mathbb{R}^n$, i.e., positive definite lattices. Recall from Sect. 1.1 that an integral lattice $L$ is a finitely generated free $\mathbb{Z}$-module provided with an integer valued symmetric bilinear form. We denote the bilinear form $L \times L \longrightarrow \mathbb{Z}$ by $(x,y) \longmapsto x \cdot y$. We also write $x^2$ for $x \cdot x$. In this section we shall study integral lattices where this form is indefinite.

Let $L$ be an indefinite integral lattice. Then $L$ contains vectors $x$ with $x^2 = 0$. A vector $x \in L$ with $x^2 = 0$ is called an *isotropic* vector of $L$. A vector $x \in L$ is called *primitive*, if $x = dy$ for $y \in L$ and $d \in \mathbb{Z}$ implies $d = \pm 1$.

**Example 4.1** We denote by 0 the $\mathbb{Z}$-module $\mathbb{Z}$ with the trivial bilinear form defined by

$$x \cdot y = 0$$

for $x, y \in \mathbb{Z}$.

**Example 4.2** Let $U$ be the free $\mathbb{Z}$-module of rank 2 generated by two elements $e_1, e_2$ provided with the symmetric bilinear form defined by

$$(x_1 e_1 + x_2 e_2) \cdot (y_1 e_1 + y_2 e_2) = x_1 y_2 + x_2 y_1.$$

Then $U$ is a unimodular integral lattice, called a *unimodular hyperbolic plane*. The matrix $((e_i \cdot e_j))$ of this form with respect to the basis $\{e_1, e_2\}$ is the matrix

$$\begin{pmatrix} 0 & 1 \\ 1 & 0 \end{pmatrix}.$$

The vectors $e_1$ and $e_2$ are primitive isotropic vectors of $U$.

Let $L$ and $L'$ be integral lattices. The *orthogonal direct sum* of $L$ and $L'$, denoted by $L \perp L'$, is the direct sum of the $\mathbb{Z}$-modules $L$ and $L'$ provided with the symmetric bilinear form defined by

$$(x + x') \cdot (y + y') = x \cdot y + x' \cdot y'$$

for $x, y \in L$ and $x', y' \in L'$.

**Example 4.3** Let $\widetilde{\Gamma}$ be a free $\mathbb{Z}$-module of rank $n+1$ with generators $e_1, \ldots, e_{n+1}$. Define the symmetric bilinear form on $\widetilde{\Gamma}$ by prescribing the values on the generators $e_1, \ldots, e_{n+1}$ by means of the Coxeter-Dynkin diagram corresponding to this set of vectors. Let the Coxeter-Dynkin diagram of the set $\{e_1, \ldots, e_n\}$ be an ordinary Coxeter-Dynkin diagram and the Coxeter-Dynkin diagram corresponding to $\{e_1, \ldots, e_{n+1}\}$ be the corresponding extended Coxeter-Dynkin diagram. This means in particular that $e_i^2 = 2$ for all $1 \leq i \leq n+1$, i.e., the generators $e_1, \ldots, e_{n+1}$ are roots in $\widetilde{\Gamma}$, and $e_1, \ldots, e_n$ generate a root lattice $\Gamma$ of rank $n$. Let

$$\beta = \sum_{i=1}^{n} m_i e_i$$

be the highest root, and define

$$z := \beta + e_{n+1}.$$

Then $z$ satisfies

$$z^2 = (e_{n+1} + \beta)^2 = 4 + 2(e_{n+1} \cdot \beta) = 0,$$
$$z \cdot e_i = 0 \quad \text{for all } 1 \le i \le n.$$

Therefore $z$ is a primitive isotropic vector in $\widetilde{\Gamma}$ which is orthogonal to $e_1, \dots, e_n$. So

$$\widetilde{\Gamma} = \Gamma \perp 0.$$

We need a lemma about primitive isotropic vectors in even unimodular lattices.

**Lemma 4.6.** *Let $z$ be a primitive isotropic vector of an even unimodular lattice $L$. Then $z$ is contained in a sublattice $U$ of $L$ which is a unimodular hyperbolic plane, and $L = U \perp L'$ for an even unimodular lattice $L'$.*

*Proof.* (a) We first show that there exists a $y \in L$ with $z \cdot y = 1$ (cf. [81, Chap. V, §3, Lemma 3]).

Let $f_z$ be the linear form $y \longmapsto z \cdot y$. Suppose $f_z$ is not surjective. Then $f_z(y) \in d\mathbb{Z}$ for some $d \in \mathbb{Z}$, $d \ge 2$ and all $y \in L$. Then $f_z = dg$ for some $g \in \mathrm{Hom}(L,\mathbb{Z})$. But $g(y) = \frac{1}{d} z \cdot y$ since $z \cdot y$ defines an isomorphism of $L$ onto its dual $L^* = \mathrm{Hom}(L,\mathbb{Z})$. This implies that $\frac{z}{d} \in L$, contradicting the fact that $z$ is primitive. Hence $f_z$ is surjective and there exists a $y \in L$ such that $y \cdot z = 1$.

(b) Choose $y \in L$ such that $y \cdot z = 1$. If $y^2 = 2m$, replace $y$ by $y - mz$ to obtain a new $y$ such that $y^2 = 0$. The submodule of $L$ generated by $\{z, y\}$ is then a unimodular hyperbolic plane $U$.

(c) Since $\mathrm{disc}(U) = -1$, one clearly has $U \cap U^\perp = \{0\}$. Moreover, if $x \in L$, the linear form $y \longmapsto x \cdot y$ $(y \in U)$ is defined by an element $x_0 \in U$. We then have $x = x_0 + x_1$ with $x_0 \in U$ and $x_1 \in U^\perp$, hence $L = U \perp U^\perp$.

(d) Since $\mathrm{disc}(L) = \pm 1$ and $\mathrm{disc}(U) = -1$, $\mathrm{disc}(U^\perp) = \pm 1$, so $U^\perp$ is unimodular.

Parts (a)-(d) constitute the proof of Lemma 4.6. □

We now embed the Leech lattice $\Lambda_{24}$ into a hyperbolic lattice. That means that we consider the lattice

$$L = \Lambda_{24} \perp U.$$

We consider coordinates $(\lambda, m, n) \in L$ with $\lambda \in \Lambda_{24}$, $m, n \in \mathbb{Z}$, and

$$(\lambda, m, n)^2 = \lambda^2 + 2mn.$$

Let $\lambda \in \Lambda_{24}$ and consider the element

$$\widetilde{\lambda} := \left(\lambda, 1, 1 - \frac{1}{2}\lambda^2\right) \in L.$$

Then $\widetilde{\lambda}$ is a root in $L$, i.e.,

$$\widetilde{\lambda}^2 = \left(\lambda, 1, 1 - \frac{1}{2}\lambda^2\right)^2 = 2.$$

Let

$$\widetilde{\Lambda} = \left\{ \widetilde{\lambda} = \left(\lambda, 1, 1 - \frac{1}{2}\lambda^2\right) \,\middle|\, \lambda \in \Lambda_{24} \right\} \subset L.$$

The set $\widetilde{\Lambda}$ can also be characterized in the following way. Let $w$ be the isotropic vector $(0,0,1) \in L$. Then $\widetilde{\Lambda}$ is the set of all roots $x$ in $L$ which satisfy

$$x \cdot w = 1.$$

For let $x = (\lambda, m, n) \in L$ be a root with $x \cdot w = 1$. Since

$$(\lambda, m, n) \cdot (0, 0, 1) = m,$$

it follows that $x \cdot w = 1$ if and only if $m$ is equal to 1. Furthermore

$$x^2 = (\lambda, 1, n)^2 = 2$$

implies that $\lambda^2 + 2n = 2$, i.e., $n = 1 - \frac{1}{2}\lambda^2$.

We identify the vectors of $\Lambda_{24}$ with the elements of the subset $\widetilde{\Lambda} \subset L$. For $\lambda, \mu \in \Lambda_{24}$ we have

$$\begin{aligned} \widetilde{\lambda} \cdot \widetilde{\mu} &= \left(\lambda, 1, 1 - \frac{1}{2}\lambda^2\right) \cdot \left(\mu, 1, 1 - \frac{1}{2}\mu^2\right) \\ &= \lambda \cdot \mu + 2 - \frac{1}{2}\lambda^2 - \frac{1}{2}\mu^2 \\ &= 2 - \frac{1}{2}(\lambda - \mu)^2. \end{aligned}$$

So we get

$$\begin{aligned} (\lambda - \mu)^2 = 4 &\iff \widetilde{\lambda} \cdot \widetilde{\mu} = 0 \\ (\lambda - \mu)^2 = 6 &\iff \widetilde{\lambda} \cdot \widetilde{\mu} = -1 \\ (\lambda - \mu)^2 = 8 &\iff \widetilde{\lambda} \cdot \widetilde{\mu} = -2 \end{aligned}$$

Let $c$ be a deep hole of $\Lambda_{24}$, and let $v_1, \ldots, v_\nu$ be the vertices of its Delaunay polytope. It follows that the hole diagram corresponding to $c$ is the Coxeter-Dynkin diagram of the system of roots $\{\widetilde{v}_1, \ldots, \widetilde{v}_\nu\}$ of the lattice $L$.

The following Lemma 4.8 is the main step in Borcherds' proof of Theorem 4.2 (cf. [5, Lemma 5.1]). For the proof of this lemma Borcherds uses an algorithm of Vinberg [90]. We modify Borcherds' proof avoiding the use of this algorithm. Instead we use the following lemma.

**Lemma 4.7.** *Let $\Gamma$ be an irreducible root lattice of rank n, and let $\widetilde{\Gamma}$ be the lattice $\Gamma \perp 0$, where the lattice $0 \subset \widetilde{\Gamma}$ is generated by a primitive element $z \in \widetilde{\Gamma}$. Let $t \in Hom(\widetilde{\Gamma}, \mathbb{Z})$ be* *a linear form with* $t(\alpha) \neq 0$ *for all roots* $\alpha$ *in* $\widetilde{\Gamma}$. *Then there exists a system* $(e_1,\ldots,e_{n+1})$ *of roots in* $\widetilde{\Gamma}$ *with the following properties:*

(i) *The Coxeter-Dynkin diagram of* $(e_1,\ldots,e_{n+1})$ *is the extended Coxeter-Dynkin diagram of* $\Gamma$.

(ii) *The primitive isotropic vector* $z$ *is equal to* $\sum_{i=1}^{n+1} m_i e_i$ *with all* $m_i > 0$ *and* $\sum_{i=1}^{n+1} m_i = h$, *where* $h$ *is the Coxeter number of* $\Gamma$.

(iii) $(e_1,\ldots,e_n)$ *is a fundamental system of roots for* $\Gamma$ *and* $\sum_{i=1}^{n} m_i e_i$ *is the highest root of* $\Gamma$.

(iv) *For all* $i$, $1 \leq i \leq n+1$, $t(e_i) > 0$.

*Proof.* Let $R$ and $\widetilde{R}$ be the sets of roots in $\Gamma$ and $\widetilde{\Gamma}$ respectively. Define

$$\widetilde{R}_t^+ = \{\alpha \in \widetilde{R} \mid t(\alpha) > 0\},\, R_t^+ = R \cap \widetilde{R}_t^+.$$

Since $t(\alpha) \neq 0$ for all $\alpha \in \widetilde{R}$, we have $\widetilde{R} = \widetilde{R}_t^+ \cup (-\widetilde{R}_t^+)$ and $R = R_t^+ \cup (-R_t^+)$. Let $\widetilde{S}_t$ be the set of indecomposable elements of $\widetilde{R}_t^+$ (cf. Sect. 1.4), and let $S_t = \widetilde{S}_t \cap R$. Then $S_t$ is a fundamental system of roots for the root lattice $\Gamma$. This follows from Lemmas 1.3, 1.4, and 1.5, where one can replace the element $t \in \mathbb{R}^n$ by a linear form $t : \mathbb{R}^n \longrightarrow \mathbb{R}$ with $t(\alpha) \neq 0$ for all $\alpha \in R$.

Let $S_t = \{e_1,\ldots,e_n\}$. The same proof as for Lemma 1.3 shows that each element of $\widetilde{R}_t^+$ is a linear combination of elements of $\widetilde{S}_t$. Since $\widetilde{R}_t^+$ generates $\widetilde{\Gamma}$, it follows that there exists an indecomposable $e_{n+1} \in \widetilde{S}_t$ not contained in $\Gamma$.

Consider the natural projection $\widetilde{\Gamma} \longrightarrow \Gamma$ and denote the image of an element $\alpha \in \widetilde{\Gamma}$ under this projection by $\overline{\alpha}$. Since $\{e_1,\ldots,e_n\}$ is a fundamental system of roots for $\Gamma$,

$$\overline{e}_{n+1} = \sum_{i=1}^{n} c_i e_i$$

with integer coefficients $c_i$, all non-negative or all non-positive. By Lemma 1.4

$$\overline{e}_{n+1} \cdot e_i \leq 0 \quad \text{for } 1 \leq i \leq n.$$

If $\overline{e}_{n+1} \in R_t^+$, then Lemma 1.5 would imply that $e_1,\ldots,e_n,\overline{e}_{n+1}$ would be linearly independent, a contradiction. Therefore $\overline{e}_{n+1} \in -R_t^+$, and all $c_i$ have to be non-positive.

We claim that $-\overline{e}_{n+1}$ is equal to the highest root $\beta$ corresponding to $\Gamma$. For $-\overline{e}_{n+1}$ satisfies

$$(-\overline{e}_{n+1}) \cdot e_i \geq 0 \quad \text{for } 1 \leq i \leq n.$$

Of course $-\overline{e}_{n+1} \leq \beta$. Therefore

$$\beta = -\overline{e}_{n+1} + \sum_{i=1}^{n} d_i e_i$$

with all $d_i \geq 0$. But the right-hand side can only be a root if all $d_i = 0$, since

$$\left(-\overline{e}_{n+1}+\sum_{i=1}^{n}d_ie_i\right)^2 = \overline{e}_{n+1}^2+2(-\overline{e}_{n+1})\cdot\left(\sum_{i=1}^{n}d_ie_i\right)+\left(\sum_{i=1}^{n}d_ie_i\right)^2$$
$$> 2$$

unless all $d_i=0$. Therefore $-\overline{e}_{n+1}=\beta$, and hence $z=\beta+e_{n+1}$. Therefore $(e_1,\ldots,e_{n+1})$ is a system of roots of $\widetilde{\Gamma}$ with the required properties. That

$$\sum_{i=1}^{m+1} m_i = h$$

follows from Lemma 1.16. This proves Lemma 4.7. □

**Lemma 4.8 (Borcherds).** *Let X be a set of elements of $\widetilde{\Lambda}$ such that the Coxeter-Dynkin diagram corresponding to X is a connected extended Coxeter-Dynkin diagram. Then X is contained in a subset $Y\subset\widetilde{\Lambda}$ such that the Coxeter-Dynkin diagram of Y is a disjoint union of extended Coxeter-Dynkin diagrams and the numbers of vertices of the corresponding ordinary Coxeter-Dynkin diagrams sum up to 24.*

*Proof.* Let $X=\{x_1,\ldots,x_{k+1}\}$, where the numbering is such that the Coxeter-Dynkin diagram of $\{x_1,\ldots,x_k\}$ is the corresponding ordinary Coxeter-Dynkin diagram. As in Example 4.3, let

$$\beta=\sum_{i=1}^{k} m_ix_i$$

be the corresponding highest root, define $m_{k+1}:=1$, and consider the vector

$$z=\sum_{i=1}^{k+1} m_ix_i.$$

Then $z$ is a primitive isotropic vector. According to Lemma 4.6, $z$ is contained in a sublattice $U'$ of $L$ which is a unimodular hyperbolic plane, $L=U'\perp N$ for an even unimodular lattice $N$ of rank 24. This lattice must be positive definite and it contains roots, e.g. the roots $x_1,\ldots,x_k$, hence it is a Niemeier lattice.

By Corollary 3.5 the irreducible components of the root sublattice of $N$ all have the same Coxeter number $h$. Let $\Gamma_i$ be such an irreducible component, different from the component spanned by $\{x_1,\ldots,x_k\}$. Consider the lattice $\widetilde{\Gamma_i}=\Gamma_i\perp\langle z\rangle$, where $\langle z\rangle$ denotes the sublattice generated by $z$. Since $w^\perp=\Lambda_{24}\perp\langle w\rangle$, there are no roots in $L$ perpendicular to $w$. So $y\longmapsto w\cdot y$ $(y\in\widetilde{\Gamma_i})$ is an integer valued linear form on $\widetilde{\Gamma_i}$ which does not vanish on the roots of $\widetilde{\Gamma_i}$. By Lemma 4.7 there exists a system $(e_1,\ldots,e_{l+1})$ of roots of $\widetilde{\Gamma_i}$ with the following properties:

(i) The Coxeter-Dynkin diagram of $(e_1,\ldots,e_{l+1})$ is the extended Coxeter-Dynkin diagram of $\widetilde{\Gamma_i}$.

(ii) $z=\sum_{j=1}^{l+1} n_je_j$, all $n_j>0$, $\sum_{j=1}^{l+1} n_j=h$.

(iii) $(e_1,\ldots,e_l)$ is a fundamental system of roots for $\Gamma$ and $\sum_{j=1}^{l} n_j e_j$ is the highest root of $\Gamma$.
(iv) For all $j$, $1 \leq j \leq l+1$, $w \cdot e_j > 0$.

We show that the elements $e_1,\ldots,e_{l+1}$ are contained in $\widetilde{\Lambda}$. Since $z = \sum_{j=1}^{k+1} m_j x_j$ and $x_j \cdot w = 1$ for $1 \leq j \leq k+1$, we have

$$h = z \cdot w.$$

Since also $z = \sum_{j=1}^{l+1} n_j e_j$, we have

$$h = \left( \sum_{j=1}^{l+1} n_j e_j \right) \cdot w = \sum_{j=1}^{l+1} n_j (e_j \cdot w).$$

By property (iv), $e_j \cdot w > 0$ for $1 \leq j \leq l+1$, so we must have $e_j \cdot w = 1$ for all $j \in \{1,\ldots,l+1\}$. Therefore $\{e_1,\ldots,e_{l+1}\} \subset \widetilde{\Lambda}$.

Now let $Y$ be the union of the set $X$ with the sets $\{e_1,\ldots,e_{l+1}\}$ for each irreducible component $\Gamma_i$ of the root sublattice of $N$ different from the component generated by $\{x_1,\ldots,x_k\}$. Then $Y$ is a subset of $\widetilde{\Lambda}$ with the required properties. This proves Lemma 4.8. □

We are now ready to complete the proof of Theorem 4.2 which was started in Sect. 4.2.

*Proof of Theorem 4.2* (cf. [5]). Let $c$ be a hole of radius greater than or equal to $\sqrt{2}$. By Proposition 4.1 and Proposition 4.2 the hole diagram of $c$ contains an extended Coxeter-Dynkin diagram. Let $V$ be the set of vertices of the Delaunay polytope $P$ of $c$ corresponding to this extended Coxeter-Dynkin diagram. By Lemma 4.8 there exists a set of vectors $W$ of the Leech lattice $\Lambda_{24}$ containing $V$, such that the Coxeter-Dynkin diagram of $W$ is a disjoint union of extended Coxeter-Dynkin diagrams and $W$ spans a sublattice of rank 24. Let $V = \{v_1,\ldots,v_{k+1}\}$, let $\widetilde{V} = \{\widetilde{v}_1,\ldots,\widetilde{v}_{k+1}\}$ be the corresponding subset of $\widetilde{\Lambda}$, and let

$$z = \sum_{i=1}^{k+1} m_i \widetilde{v}_i$$

be the corresponding primitive isotropic vector as in the proof of Lemma 4.8. Then $z$ can be written as follows in the coordinates $(\lambda,m,n)$ of $L = \Lambda_{24} \perp U$:

$$z = \left( \zeta, h, -\frac{1}{2}\frac{\zeta^2}{h} \right),$$

where

$$\zeta = \sum_{i=1}^{k+1} m_i v_i \in \Lambda_{24}$$

and $h$ is the Coxeter number corresponding to $\widetilde{v}_1,\ldots,\widetilde{v}_{k+1}$. This follows from Lemma 1.16. Now for $\widetilde{\lambda} = (\lambda, 1, 1 - \frac{1}{2}\lambda^2) \in \widetilde{\Lambda}$ we have

$$
\begin{aligned}
& z\cdot\widetilde{\lambda}=0 \\
\iff & \left(\zeta,h,-\frac{1}{2}\frac{\zeta^2}{h}\right)\cdot\left(\lambda,1,1-\frac{1}{2}\lambda^2\right)=0 \\
\iff & \zeta\cdot\lambda+h\left(1-\frac{1}{2}\lambda^2\right)-\frac{1}{2}\frac{\zeta^2}{h}=0 \\
\iff & h\left(1-\frac{1}{2}\left(\frac{\zeta^2}{h^2}-2\frac{\zeta}{h}\cdot\lambda+\lambda^2\right)\right)=0 \\
\iff & 1-\frac{1}{2}\left(\frac{\zeta}{h}-\lambda\right)^2=0 \\
\iff & \left(\frac{\zeta}{h}-\lambda\right)^2=2.
\end{aligned}
$$

Therefore $c'=\frac{\zeta}{h}$ is a vector of $\Lambda_{24}\otimes\mathbb{R}$ which has distance $\sqrt{2}$ from the points of $V$ and is the centre of a hole of radius $\sqrt{2}$. The vectors of $W$ are the vertices of the Delaunay polytope $P'$ of $c'$. The point $c'$ is also the centre of $V$. But since $V$ corresponds to a component of the hole diagram, we have $(v-w)^2=4$ for all vertices $v,w$ of $P$ with $v\in V$, $w\notin V$. Therefore the centre of $V$ is also the centre of $P$ (cf. Fig. 4.3). Hence we must have

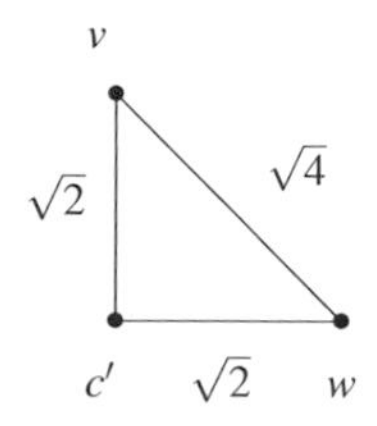


**Fig. 4.3** Two non-joined vertices in the hole diagram

$P=P'$ and $c=c'$. Hence $c$ has radius $\sqrt{2}$, and $\Lambda$ has covering radius $\sqrt{2}$. This proves Theorem 4.2. □

For the proof of Theorem 4.3 we need some auxiliary results.

Let $N$ be a Niemeier lattice with Weyl vector $\rho$ and Coxeter number $h$. We consider the lattice

$$M=N\perp U.$$

Let $(y,m,n)$ be coordinates for $M$ with $y\in N$, $m,n\in\mathbb{Z}$ such that

$$(y,m,n)^2=y^2+2mn.$$

By Lemma 4.4 the vector $w:=(\rho,h,-(h+1))$ is in $M$ and by Lemma 4.3 it is isotropic. Lemma 4.6 implies that $w$ is contained in a sublattice $U'$ of $M$ which is a unimodular hyperbolic plane, and $M=U'\perp N'$ for an even unimodular lattice $N'$. We will show that the lattice $N'$ has at most half as many roots as $N$ (cf. [5, Lemma 4.2]. In fact it follows from Sect. 4.3 that $N'$ has no roots and is isomorphic to the Leech lattice. Unfortunately we don't know of any elementary direct proof of this fact. We can only show that there are no roots in $N'$ which have inner product 0 or $\pm 1$ with $z = (0,0,1)$ (cf. [5, Lemma 4.1]).

**Lemma 4.9.** *There are no roots of $M$ that are perpendicular to $w$ and have inner product 0 or $\pm 1$ with $z = (0,0,1)$.*

*Proof.* Let $r = (y,m,n)$ be a root of $M$ that is perpendicular to $w$. First suppose that $r$ has inner product 0 with $z$. Then $r \cdot z = 0$ implies $m = 0$, and hence $r^2 = y^2 = 2$, so $y$ is a root of $N$. Furthermore

$$0 = r \cdot w = (y,0,n) \cdot (\rho, h, -(h+1)) = y \cdot \rho + nh,$$

and thus $y \cdot \rho = -nh$. But this is impossible, since we have $1 \leq |y \cdot \rho| \leq h-1$ for any root $y$ of $N$.

Now suppose that $r$ has inner product 1 with $z$. Then $r \cdot z = 1$ implies $m = 1$. Furthermore

$$0 = r \cdot w = (y,1,n) \cdot (\rho, h, -(h+1)) = y \cdot \rho - (h+1) + nh,$$

and

$$2 = r^2 = (y,1,n)^2 = y^2 + 2n.$$

Therefore

$$\begin{aligned}\left(y - \frac{\rho}{h}\right)^2 &= y^2 - 2\frac{y \cdot \rho}{h} + \frac{\rho^2}{h^2} \\ &= 2 - 2n + 2\frac{nh-h-1}{h} + 2(h+1)h \\ &= 2.\end{aligned}$$

But this is impossible by Lemma 4.5.

If $r$ has inner product $-1$ with $z$, then $-r$ has inner product 1 with $z$. But we have just seen that this is impossible. Hence no root in $w^\perp$ can have inner product 0 or $\pm 1$ with $z$. This proves Lemma 4.9. □

The following lemma is [5, Lemma 4.2].

**Lemma 4.10.** *If $N'$ contains roots, i.e., if $N'$ is a Niemeier lattice, then its Coxeter number $h'$ is at most $\frac{1}{2}h$.*

*Proof.* Assume that $N'$ contains roots. Let $\Gamma_i$ be any irreducible component of the root sublattice of $N'$. By Lemma 4.9 all roots in $w^\perp$ have nonzero inner product with $z$. Therefore it follows from Lemma 4.7 that we can find a system $(e_1, \ldots, e_{n+1})$ of roots in $w^\perp$ with the following properties:

(i) $(e_1, \ldots, e_{n+1})$ is a fundamental system of roots for $\Gamma_i$.

(ii) $w = \sum_{j=1}^{n+1} m_j e_j$ with all $m_j > 0$, and $\sum_{j=1}^{n+1} m_j = h'$.

(iii) For all $j$, $1 \leq j \leq n+1$, $e_j \cdot z > 0$.

Now

$$\left(\sum_{j=1}^{n+1} m_j e_j\right) \cdot z = (\rho, h, -(h+1)) \cdot (0,0,1) = h.$$

Since $e_j \cdot z > 0$, Lemma 4.9 implies that $e_j \cdot z \geq 2$. Since this is true for all $j$, $1 \leq j \leq n+1$, we obtain $h' \leq \frac{1}{2}h$ and therefore the claim of Lemma 4.10. □

**Lemma 4.11.** *For any Niemeier lattice $N$, the lattice $M = N \perp U$ can also be decomposed as $M = \Lambda \perp U'$, where $U'$ is another unimodular hyperbolic plane and $\Lambda$ is isomorphic to the Leech lattice.*

*Proof.* By Lemma 4.10 the lattice $M = N \perp U$ can also be decomposed as $M = N' \perp U'$, where $U'$ is a unimodular hyperbolic plane and $N'$ a Niemeier lattice with Coxeter number at most $\frac{1}{2}h$ or a lattice containing no roots. By repeating this we eventually get an even unimodular 24-dimensional lattice $\Lambda$ with no roots. By the uniqueness of the Leech lattice, this must be the Leech lattice. Therefore Lemma 4.11 is proved. □

Lemma 4.11 also follows as a special case from a general structure theorem about indefinite unimodular integral lattices: see [81, Chap. V, §2, Theorem 6].

*Proof of Theorem 4.3.* Let $c$ be a deep hole of $\Lambda_{24}$. Then the hole diagram of $c$ contains an extended Coxeter-Dynkin diagram. Let $x_1, \ldots, x_{k+1}$ be the vectors of $\widetilde{\Lambda}$ corresponding to it and let

$$z = \sum_{i=1}^{k+1} m_i x_i$$

be the corresponding primitive isotropic vector. As we saw in the proof of Theorem 4.2, $z$ can be written as follows in the coordinates $(\lambda, m, n)$ of $L = \Lambda_{24} \perp U$:

$$z = \left(\zeta, h, -\frac{1}{2}\frac{\zeta^2}{h}\right),$$

where $\zeta$ is a certain vector in $\Lambda_{24}$ and $h$ is the Coxeter number corresponding to $x_1, \ldots, x_{k+1}$. As we have shown in the course of the proof of Lemma 4.8, $z^\perp$ contains a Niemeier lattice $N$ with Coxeter number $h$. Let $v_1, \ldots, v_\nu$ be the vertices of the Delaunay polytope of $c$, and let $\widetilde{v}_1, \ldots, \widetilde{v}_\nu$, $\widetilde{v}_i = (v_i, 1, 1 - \frac{1}{2}v_i^2)$, be the corresponding vectors of $\widetilde{\Lambda}$. Now for $\widetilde{\lambda} = (\lambda, 1, 1 - \frac{1}{2}\lambda^2) \in \widetilde{\Lambda}$ we have

$$z \cdot \widetilde{\lambda} = 0 \Leftrightarrow \left(\frac{\zeta}{h} - \lambda\right)^2 = 2,$$

as we saw in the proof of Theorem 4.2. Since $c$ is the unique vector in $\Lambda_{24} \otimes \mathbb{R}$ which has distance $\sqrt{2}$ from $v_1, \ldots, v_\nu$, it follows that $c = \frac{\zeta}{h}$ and $z \cdot \widetilde{v}_i = 0$ for all $i$, $1 \leq i \leq \nu$. Therefore the vectors $\widetilde{v}_1, \ldots, \widetilde{v}_\nu$ are roots in the Niemeier lattice $N$. Since

$$\widetilde{v}_i \cdot \widetilde{v}_j = \left(v_i, 1, 1 - \frac{1}{2}v_i^2\right) \cdot \left(v_j, 1, 1 - \frac{1}{2}v_j^2\right)$$

$$\begin{aligned} &= v_i \cdot v_j + 1 - \frac{1}{2}v_i^2 + 1 - \frac{1}{2}v_j^2 \\ &= 2 - \frac{1}{2}(v_i - v_j)^2 \\ &= (v_i - c)\cdot(v_j - c), \end{aligned}$$

the hole diagram of $c$ is the Coxeter-Dynkin diagram corresponding to the vectors $\widetilde{v}_1, \ldots, \widetilde{v}_\nu$. Therefore the components of the hole diagram of $c$ are the extended Coxeter-Dynkin diagrams of the irreducible components of the root sublattice of $N$. Since $N$ is determined up to isomorphism by its root sublattice, this establishes a mapping from the set of deep holes of $\Lambda_{24}$ to the set of isomorphism classes of even unimodular lattices in $\mathbb{R}^{24}$ containing roots. It is clear that the isomorphism class of $N$ only depends on the congruence class of $c$ under congruences of $\Lambda_{24}$, because the congruence class of $c$ is determined by the hole diagram. It also follows that non-equivalent deep holes correspond to non-isomorphic Niemeier lattices.

It remains to show the surjectivity of this mapping. But by Lemma 4.11, for any Niemeier lattice $N$ we can find a vector $z \in L = \Lambda_{24} \perp U$, such that $z^\perp$ contains a lattice isomorphic to $N$. This proves the surjectivity of the mapping and completes the proof of Theorem 4.3. □

Conway and Sloane have found nice coordinates for the lattice $L = \Lambda \perp U$ which has led to very simple definitions of the Leech lattice, see [19] (= [21, Chap. 26]). We shall indicate the most elegant of these constructions.

Let $\mathbb{R}^{n,1}$ denote the real $(n+1)$-dimensional vector space of vectors

$$x = (x_1, \ldots, x_n \,|\, x_{n+1})$$

with inner product defined by

$$x \cdot y = x_1 y_1 + \ldots + x_n y_n - x_{n+1} y_{n+1}$$

for $x = (x_1, \ldots, x_n \,|\, x_{n+1})$, $y = (y_1, \ldots, y_n \,|\, y_{n+1}) \in \mathbb{R}^{n,1}$. Assume that $n \equiv 1 \pmod 8$. Define $\mathrm{I}_{n,1}$ to be the set of vectors $x = (x_1, \ldots, x_n \,|\, x_{n+1})$ with all $x_i \in \mathbb{Z}$. This is an odd unimodular lattice. Let $\Gamma$ be the submodule of $\mathrm{I}_{n,1}$ formed of elements $x$ such that $x \cdot x$ is even, i.e.,

$$x_1 + \ldots + x_n - x_{n+1} \in 2\mathbb{Z}.$$

One has $[\mathrm{I}_{n,1} : \Gamma] = 2$. Let $\mathrm{II}_{n,1}$ be the lattice in $\mathbb{R}^{n,1}$ generated by $\Gamma$ and by

$$u = \left(\frac{1}{2}, \ldots, \frac{1}{2} \,\middle|\, \frac{1}{2}\right).$$

Since $n \equiv 1 \pmod 8$, one has $2u \in \Gamma$. But $u \notin \Gamma$, so $[\mathrm{II}_{n,1} : \Gamma] = 2$. The lattice $\mathrm{II}_{n,1}$ is the set of all $x \in \mathbb{R}^{n,1}$ for which all $x_i$ are all in $\mathbb{Z}$ or all in $\mathbb{Z} + \frac{1}{2}$ and which satisfy

$$x_1 + \ldots + x_n - x_{n+1} \in 2\mathbb{Z}.$$

Therefore we have $x \cdot u = \frac{1}{2}(x_1 + \ldots x_n - x_{n+1}) \in \mathbb{Z}$. Since $u \cdot u = \frac{n-1}{4}$, this shows that $\mathrm{II}_{n,1}$ is an even integral lattice. Moreover the fact that $\Gamma$ has the same index in $\mathrm{I}_{n,1}$ as in $\mathrm{II}_{n,1}$ shows that $\mathrm{II}_{n,1}$ is unimodular, too. ($\mathrm{II}_{n,1}$ is a neighbour of $\mathrm{I}_{n,1}$, cf. Sect. 4.1.)

The lattice $\mathrm{II}_{n,1}$ is the unique even unimodular hyperbolic lattice of rank $n+1$ up to isomorphism. For $n = 25$ this follows from Lemma 4.6 and Lemma 4.11, for the general case see [81, Chap. V, §2, Theorem 6].

Now consider the lattice $\mathrm{II}_{25,1}$ and the vector

$$w = (0, 1, 2, 3, \ldots, 23, 24 \,|\, 70)$$

in $\mathrm{II}_{n,1}$. It is known that the only solutions of

$$0^2 + 1^2 + 2^2 + \ldots + m^2 = n^2$$

in positive integers are $(m, n) = (1, 1)$ and $(m, n) = (24, 70)$ (cf. [65, p. 258]). Therefore $w$ is an isotropic vector in $\mathrm{II}_{25,1}$. Conway and Sloane [19] proved the following theorem by using Theorem 4.4.

**Theorem 4.5 (Conway, Sloane).** *The lattice $w^{\perp}/\langle w \rangle$ is isomorphic to the Leech lattice.*

For the proof we refer to [19].

Similarly, consider the lattice $\mathrm{II}_{9,1}$ and the isotropic vector

$$w = \frac{1}{2}(1, 1, 1, 1, 1, 1, 1, 1, 1 \,|\, 3)$$

in $\mathrm{II}_{9,1}$. Then $w^{\perp}/\langle w \rangle$ is isomorphic to the lattice $E_8$ (cf. [14], [70], and [19, Lemma 6]).

**Exercise 4.1** Find the roots in $w^{\perp}/\langle w \rangle$.

**Exercise 4.2** Consider the lattice $\mathrm{I}_{n,1}$ for $1 \leq n \leq 8$. Let

$$w := (1, \ldots, 1 | 3) \in \mathrm{I}_{n,1}.$$

Show that the lattice $w^{\perp} \subset \mathrm{I}_{n,1}$ is an even integral positive definite lattice of rank $n$ and discriminant $-w^2$.

**Exercise 4.3** Derive Table 4.1 of lattices $w^{\perp}$ for $1 \leq n \leq 8$. Here the ring $\mathfrak{O}$ is the subring $\{\frac{a}{2} + \frac{b}{2}\sqrt{-7} \,|\, a, b \in \mathbb{Z},\, a \equiv b (\mathrm{mod}\ 2)\} \subset \mathbb{C} \cong \mathbb{R}^2$ with the symmetric bilinear form defined in Sect. 5.1, and $\langle 2\sqrt{2} \rangle$ denotes the lattice in $\mathbb{R}$ spanned by $2\sqrt{2}$.

## 4.5 Automorphism Groups

In this section we state Conway's remarkable result [13] (= [21, Chap. 27]) on the relation between the automorphism group of the Leech lattice and the automorphism group of the 26-dimensional unimodular hyperbolic lattice.

**Table 4.1** Lattices $w^\perp \subset \mathrm{I}_{n,1}$, $1 \leq n \leq 8$

| $n$ | $w^2$ | $w^\perp$ |
|---|---|---|
| 8 | $-1$ | $E_8$ |
| 7 | $-2$ | $E_7$ |
| 6 | $-3$ | $E_6$ |
| 5 | $-4$ | $D_5$ |
| 4 | $-5$ | $A_4$ |
| 3 | $-6$ | $A_1 \perp A_2$ |
| 2 | $-7$ | ring $\mathfrak{O}$ of integers of $\mathbb{Q}(\sqrt{-7})$ |
| 1 | $-8$ | $\langle 2\sqrt{2} \rangle$ |

As in Sect. 4.4, let $L = \Lambda_{24} \perp U$ and let $\mathrm{Aut}(L)$ denote the automorphism group of $L$. As above, we consider coordinates $(\lambda, m, n) \in L$ with $\lambda \in \Lambda_{24}$, $m, n \in \mathbb{Z}$, and

$$(\lambda, m, n)^2 = \lambda^2 + 2mn.$$

Let $w = (0,0,1)$ and

$$\begin{aligned} \widetilde{\Lambda} &= \left\{ \widetilde{\lambda} = \left( \lambda, 1, 1 - \frac{1}{2}\lambda^2 \right) \,\middle|\, \lambda \in \Lambda_{24} \right\} \\ &= \{ x \in L \,|\, x^2 = 2 \text{ and } x \cdot w = 1 \}. \end{aligned}$$

We denote by $R$ the subgroup of $\mathrm{Aut}(L)$ generated by the reflections corresponding to the elements of $\widetilde{\Lambda}$ and $\pm 1$.

The automorphism group $Co_0 := \mathrm{Aut}(\Lambda_{24})$ of the Leech lattice is a group of order

$$2^{22} 3^9 5^4 7^2 11 \cdot 13 \cdot 23 = 8315553613086720000.$$

The quotient group $Co_1 := Co_0 / \{\pm 1\}$ is a famous sporadic simple group discovered by Conway [11]. We define a group $Co_\infty$ as follows: Let $\Lambda_{24}$ act on $\Lambda_{24}$ by translations. Then $Co_\infty$ is the semi-direct product

$$Co_\infty := \Lambda_{24} \cdot \mathrm{Aut}(\Lambda_{24}).$$

This is the affine automorphism group of the Leech lattice.

Let $G$ be a group acting on a set $X$ with group action $\phi : G \times X \to X$. The *stabilizer* of a point $x \in X$ is the subgroup $\mathrm{Stab}(x) = \{ g \in G \,|\, \phi(g, x) = x \}$. One says that $G$ acts *simply transitively* on $X$ if it acts transitively and the stabilizers of all the elements $x \in X$ are trivial.

Conway has proved [13] (= [21, Chap. 27]) (see also [5]):

**Theorem 4.6 (Conway).** *The automorphism group* $\mathrm{Aut}(L)$ *of the lattice* $L = \Lambda_{24} \perp U$ *is the semi-direct product of the subgroups* $R$ *and* $Co_\infty$.

*Proof.* a) We first show that $R$ acts transitively on the primitive isotropic vectors of $L$ that are not orthogonal to any root of $L$. Let $z = (\xi, a, b)$ be such a vector where $\xi \in \Lambda_{24}$,

$0 \neq a \in \mathbb{Z}$, and $b = -\frac{1}{2}\frac{\xi^2}{a}$. For $\widetilde{\lambda} = (\lambda, 1, 1 - \frac{1}{2}\lambda^2) \in \widetilde{\Lambda}$ we have

$$z \cdot \widetilde{\lambda} = a - \frac{1}{2}a\left(\frac{\xi}{a} - \lambda\right)^2.$$

Since $z$ is not orthogonal to any root, in particular $z \cdot \widetilde{\lambda} \neq 0$ for all $\widetilde{\lambda} \in \widetilde{\Lambda}$. This implies

$$\left(\frac{\xi}{a} - \lambda\right)^2 \neq 2$$

for all $\widetilde{\lambda} \in \widetilde{\Lambda}$. By Theorem 4.3 the covering radius of the Leech lattice is $\sqrt{2}$. Therefore $\frac{\xi}{a}$ cannot be the centre of a deep hole of $\Lambda_{24}$ and there exists a $\widetilde{\lambda} = (\lambda, 1, 1 - \frac{1}{2}\lambda^2) \in \widetilde{\Lambda}$ with

$$\left(\frac{\xi}{a} - \lambda\right)^2 < 2.$$

Fix such a $\widetilde{\lambda}$. Let

$$a' := \frac{1}{2}a\left(\frac{\xi}{a} - \lambda\right)^2.$$

Then $0 \leq |a'| < |a|$ and

$$s_{\widetilde{\lambda}}(z) = z - (z \cdot \widetilde{\lambda})\widetilde{\lambda} = \left(\xi - (a - a')\lambda, a', -\frac{1}{2}\frac{\xi^2}{a} - (a - a')\left(1 - \frac{1}{2}\lambda^2\right)\right).$$

If $a' = 0$ then $\xi - a\lambda = 0$. This means that

$$s_{\widetilde{\lambda}}(z) = (0, 0, \pm 1).$$

Otherwise, let

$$b' := -\frac{1}{2}\frac{\xi^2}{a} - (a - a')\left(1 - \frac{1}{2}\lambda^2\right).$$

Since $(s_{\widetilde{\lambda}}(z))^2 = 0$, we have

$$-2a'b' = (\xi - (a - a')\lambda)^2.$$

Therefore $2a'b' \leq 0$. Suppose $a > 0$. Then $a' > 0$ and hence $b' \leq 0$. Since $b = -\frac{1}{2}\frac{\xi^2}{a} < 0$, $a - a' > 0$, and $1 - \frac{1}{2}\lambda^2 < 0$ if $\lambda \neq 0$, we get $0 \leq |b'| < |b|$. The same is true when $a < 0$.

If $b' = 0$ then $\xi - (a - a')\lambda = 0$, and hence

$$s_{\widetilde{\lambda}}(z) = (0, \pm 1, 0).$$

The vector $s_{\widetilde{\lambda}}(z)$ is again not orthogonal to any root. For let $x \in L$ be a root. Then $s_{\widetilde{\lambda}}(z) \cdot x = z \cdot s_{\widetilde{\lambda}}(x) \neq 0$ since $s_{\widetilde{\lambda}}(x)$ is again a root.

If $a' \neq 0$ and $b' \neq 0$ we can repeat the process with $a := a'$ and $b := b'$. Since $0 \leq |a'| < |a|$ and $0 \leq |b'| < |b|$, after finitely many steps we will have $a' = 0$ or $b' = 0$. Therefore we see that $z$ is equivalent under the action of $R$ to $(0, \pm 1, 0)$ or to $(0, 0, \pm 1)$. For $\widetilde{0} = (0,1,1) \in \widetilde{\Lambda}$ we have

$$s_{\widetilde{0}}(0,1,0) = (0,0,-1).$$

By definition, $R$ contains $\pm 1$. This shows that $z$ is equivalent under the action of $R$ to $w = (0,0,1)$. Moreover, since $z \cdot \widetilde{\lambda} \neq 0$ for every $\widetilde{\lambda} \in \widetilde{\Lambda}$, we see that there exists a unique element $r \in R$ with $r(z) = w$. This means that $R$ acts simply transitively on the set of primitive isotropic vectors of $L$ which are not orthogonal to any roots.

b) Now let $z \in L$ again be a primitive isotropic vector that is not orthogonal to any root and let $\mathrm{Stab}(z) \subset \mathrm{Aut}(L)$ be the stabilizer of $z$. We show that $\mathrm{Stab}(z)$ is isomorphic to $Co_\infty$. An automorphism of $L$ which fixes $z$ must also map the orthogonal complement of $z$ to itself. Since the orthogonal complement does not contain any root, it is isomorphic to

$$\Lambda_{24} \perp 0.$$

Now an automorphism $g$ of $\Lambda_{24} \perp 0$ which fixes the generator $z$ of the lattice 0 must be of the form

$$g(\lambda + az) = h(z) + \phi(\lambda)z + az \qquad (\lambda \in \Lambda_{24}, a \in \mathbb{Z})$$

for some $h \in \mathrm{Aut}(\Lambda_{24})$ and $\phi \in \Lambda_{24}^*$. Since $\Lambda_{24}$ is unimodular, $\phi$ is of the form $f_x$ for an element $x \in \Lambda_{24}$ where $f_x$ is defined by $f_x(y) = x \cdot y$ for $y \in \Lambda_{24}$. Hence $\mathrm{Stab}(z)$ is the semi-direct product of $\mathrm{Aut}(\Lambda_{24})$ and $\Lambda_{24}$ where $x \in \Lambda_{24}$ acts on $\Lambda_{24} \perp 0$ by sending $\lambda + az$ to $\lambda + (x \cdot \lambda)z + az$. Hence $\mathrm{Stab}(z)$ is isomorphic to $Co_\infty$. This proves Theorem 4.6. □

**Remark 4.2** Let $\Gamma$ be an even unimodular 24-dimensional lattice without any roots. By the general structure theorem about indefinite unimodular integral lattices [81, Chap. V, §2, Theorem 6], the lattices $\Gamma \perp U$ and $L = \Lambda \perp U$ are isomorphic. By Sect. 4.4, the lattices $\Gamma$ and $\Lambda$ correspond to primitive isotropic vectors of $L$ which are not orthogonal to any root of $L$. By part (a) of the proof of Theorem 4.6, there is an element of $\mathrm{Aut}(L)$ which maps these vectors onto each other and hence maps $\Gamma$ onto $\Lambda$. This shows again that the Leech lattice is unique (cf. [5, Corollary 6.2]).

**Exercise 4.4** The covering radius of the lattice $E_8$ is 1. Using this fact, prove that the automorphism group $\mathrm{Aut}(L)$ of the lattice $L = E_8 \perp U$ is generated by the reflections corresponding to the roots of $L$ and $\pm 1$. (Hint: This can be proved along the same lines as Theorem 4.6.)

For further results concerning the embedding of the Leech lattice and other definite lattices in a hyperbolic lattice see [18] (= [21, Chap. 28]) and [70].

The Leech lattice occurs in many different connections. For example, the Leech lattice is related to the Fischer-Griess "Monster" simple group (see [21, Chap. 29] and [15]). There were discovered some rather mysterious connections between automorphisms of the Leech lattice and singularities of complex surfaces (see [77] and [22]).

# Chapter 5
# Lattices over Integers of Number Fields and Self-Dual Codes

## 5.1 Lattices over Integers of Cyclotomic Fields

Up to now, we have only considered binary codes. In this chapter we want to present a generalization of the results of Chapter 2 to self-dual codes over $\mathbb{F}_p$, where $p$ is an odd prime number. The results which we want to discuss are due to G. van der Geer and F. Hirzebruch [36, pp. 759-798]. In Sect. 1.3 we associated an integral lattice in $\mathbb{R}^n$ to a binary linear code of length $n$. In Sect. 5.2 we shall generalize this construction by associating a lattice over the integers of a cyclotomic field to a linear code over $\mathbb{F}_p$. In this section we shall study lattices over integers of cyclotomic fields. For the background on algebraic number theory see also [76] and [87].

Let $p$ be an odd prime number. Let $\zeta := e^{2\pi i/p}$. Then $\zeta^r$ for $r = 0, 1, \ldots, p-1$ are the $p$-th roots of unity, and

$$x^p - 1 = (x-1)(x-\zeta)\cdots(x-\zeta^{p-1}).$$

The polynomial

$$1 + x + x^2 + \ldots + x^{p-1} = (x^p - 1)/(x-1)$$

is irreducible over $\mathbb{Q}$. Its roots in $\mathbb{C}$ are the primitive $p$-th roots of unity $\zeta, \zeta^2, \ldots, \zeta^{p-1}$, i.e., over $\mathbb{C}$

$$1 + x + x^2 + \ldots + x^{p-1} = (x-\zeta)\cdots(x-\zeta^{p-1}).$$

By setting $x = 1$ in this equation, one derives the equation

$$p = (1-\zeta)(1-\zeta^2)\cdots(1-\zeta^{p-1}). \tag{5.1}$$

We are going to study the field $K := \mathbb{Q}(\zeta)$. It is the field obtained by adjoining $\zeta$ to $\mathbb{Q}$. Since $p-1$ is the degree of the minimal polynomial of $\zeta$ over $\mathbb{Q}$, $K$ is a vector space over $\mathbb{Q}$ of dimension $p-1$. The elements $1, \zeta, \ldots, \zeta^{p-2}$ form a basis of this vector space, so each element $\alpha \in \mathbb{Q}(\zeta)$ has a unique representation

$$\alpha = a_0 + a_1\zeta + \ldots + a_{p-2}\zeta^{p-2}, \qquad a_i \in \mathbb{Q}.$$

We shall show that the set

$$\left\{ \alpha = \sum_{i=0}^{p-2} a_i \zeta^i \;\middle|\; a_i \in \mathbb{Z} \quad \text{for } i = 0, 1, \dots, p-2 \right\}$$

is the ring of integers of $K$. For this purpose we need to calculate the traces of some elements in $K$.

Let $K$ be an *algebraic number field*, i.e., a field extension of $\mathbb{Q}$ of finite degree $n = [K : \mathbb{Q}]$.

An element $x \in K$ is called *integral* over $\mathbb{Z}$, if there exist $a_0, \dots, a_{m-1} \in \mathbb{Z}$ such that

$$a_0 + a_1 x + \dots + a_{m-1} x^{m-1} + x^m = 0,$$

i.e., if $x$ is a root of a monic polynomial with integral coefficients. The elements of a number field $K$ which are integral over $\mathbb{Z}$ are called the *integers* of $K$. They form a subring $\mathfrak{O}$ of $K$. This ring $\mathfrak{O}$ is a free $\mathbb{Z}$-module of rank $n$.

For a number field $K$ of degree $n$, there exist $n$ distinct embeddings $\sigma_i : K \to \mathbb{C}$, $i = 1, \dots, n$. These embeddings $\sigma_i$ fix $\mathbb{Q}$, i.e., $\sigma_i(\alpha) = \alpha$ for all $\alpha \in \mathbb{Q}$. Then the *trace* of an element $\alpha \in K$ over $\mathbb{Q}$, denoted by $\mathrm{Tr}_{K/\mathbb{Q}}(\alpha)$, is defined by

$$\mathrm{Tr}_{K/\mathbb{Q}}(\alpha) := \sum_{i=1}^{n} \sigma_i(\alpha).$$

One has $\mathrm{Tr}_{K/\mathbb{Q}}(\alpha) \in \mathbb{Q}$. The number $\mathrm{Tr}_{K/\mathbb{Q}}(\alpha)$ is the trace of the endomorphism of the $\mathbb{Q}$-vectorspace $K$ defined by multiplication by $\alpha$. If there is no danger of confusion, we write $\mathrm{Tr}(\alpha)$ instead of $\mathrm{Tr}_{K/\mathbb{Q}}(\alpha)$.

For $K = \mathbb{Q}(\zeta)$ there are $p-1$ embeddings $\sigma_1, \dots, \sigma_{p-1} : K \to \mathbb{C}$ given by

$$\sigma_r(\zeta) = \zeta^r, \quad r = 1, \dots, p-1.$$

Then the trace of an element $\alpha \in \mathbb{Q}(\zeta)$, $\alpha = a_0 + a_1\zeta + \dots + a_{p-2}\zeta^{p-2}$, is easily computed to be

$$\mathrm{Tr}(\alpha) = (p-1)a_0 - a_1 - a_2 - \dots - a_{p-2}. \tag{5.2}$$

Note that the $\sigma_r$ leave $\mathbb{Q}(\zeta)$ invariant. Thus they form a group (with respect to composition), the Galois group of $\mathbb{Q}(\zeta)$ over $\mathbb{Q}$.

Let $\mathfrak{O}$ be the ring of integers of $\mathbb{Q}(\zeta)$. Evidently $\mathfrak{O}$ contains $\zeta$ and its powers. Let $\mathfrak{P} := (1-\zeta)$ be the principal ideal of $\mathfrak{O}$ generated by the element $1 - \zeta \in \mathfrak{O}$.

**Lemma 5.1.** $\mathfrak{P} \cap \mathbb{Z} = p\mathbb{Z}$.

*Proof.* Formula (1) shows that $p \in \mathfrak{P}$. Thus $p\mathbb{Z} \subset \mathfrak{P} \cap \mathbb{Z}$. Since $p\mathbb{Z}$ is a maximal ideal of $\mathbb{Z}$, the assumption $\mathfrak{P} \cap \mathbb{Z} \neq p\mathbb{Z}$ implies $\mathfrak{P} \cap \mathbb{Z} = \mathbb{Z}$, i.e., that $1 - \zeta$ is a unit in $\mathfrak{O}$. But in this case the conjugates $(1 - \zeta^i)$ of $1 - \zeta$ must also be units; hence $p$ must be a unit in $\mathfrak{O} \cap \mathbb{Z}$ by formula (5.1); and thus $p^{-1}$ must belong to $\mathbb{Z}$, which is absurd. This proves Lemma 5.1. □

**Lemma 5.2.** *For any* $\alpha \in \mathfrak{P}$, $\mathrm{Tr}(\alpha) \in p\mathbb{Z}$.

*Proof.* Let $\alpha \in \mathfrak{P}$. Then $\alpha$ can be written $\alpha = y(1-\zeta)$ for some $y \in \mathfrak{O}$. Each conjugate $y_i(1-\zeta^i)$ of $y(1-\zeta)$ is a multiple (in $\mathfrak{O}$) of $1-\zeta^i$, which is itself a multiple of $1-\zeta$, since $1-\zeta^i = (1-\zeta)\left(1+\zeta+\ldots+\zeta^{i-1}\right)$. Since the trace is the sum of the conjugates, we have $\operatorname{Tr}(y(1-\zeta)) \in \mathfrak{P}$. Lemma 5.2 now follows immediately from Lemma 5.1, for the trace of an integer belongs to $\mathbb{Z}$. $\square$

**Proposition 5.1.** *The ring $\mathfrak{O}$ of integers of the cyclotomic field $\mathbb{Q}(\zeta)$ is the subring*

$$\left\{ \alpha = \sum_{i=0}^{p-2} a_i \zeta^i \;\middle|\; a_i \in \mathbb{Z} \quad \textit{for } i = 0, \ldots, p-2 \right\} \subset K$$

*of $K$.*

*Proof.* Let $\alpha = a_0 + a_1\zeta + \ldots + a_{p-2}\zeta^{p-2}$, $a_i \in \mathbb{Q}$, be an element of $\mathfrak{O}$. Then

$$\alpha(1-\zeta) = a_0(1-\zeta) + a_1(\zeta-\zeta^2) + \ldots + a_{p-2}\left(\zeta^{p-2} - \zeta^{p-1}\right).$$

Therefore $\operatorname{Tr}(\alpha(1-\zeta)) = a_0 \operatorname{Tr}(1-\zeta) = a_0 p$. By Lemma 5.2, $pa_0 \in p\mathbb{Z}$, so $a_0 \in \mathbb{Z}$. Since $\zeta^{-1} = \zeta^{p-1}$, one has $\zeta^{-1} \in \mathfrak{O}$. Therefore

$$(\alpha - a_0)\zeta^{-1} = a_1 + a_2\zeta + \ldots + a_{p-2}\zeta^{p-3} \in \mathfrak{O}.$$

By the same argument as before, $a_1 \in \mathbb{Z}$. Applying the same argument successively, we conclude that each $a_i \in \mathbb{Z}$. This proves Proposition 5.1. $\square$

We continue the study of the principal ideal $\mathfrak{P}$ of $\mathfrak{O}$.

**Lemma 5.3.** $\mathfrak{P}^{p-1} = (p) \subset \mathfrak{O}$.

*Proof.* Consider the element

$$\frac{1-\zeta^r}{1-\zeta^s} \in \mathbb{Q}(\zeta), \quad r \equiv ts \pmod{p}.$$

Since $1-\zeta^i = (1-\zeta)(1+\zeta+\ldots+\zeta^{i-1})$, we have

$$\frac{1-\zeta^r}{1-\zeta^s} = \frac{1-\zeta^{ts}}{1-\zeta^s} = 1 + \zeta^s + \ldots + \zeta^{(t-1)s} \in \mathfrak{O}.$$

We conclude that $\mathfrak{P} = (1-\zeta^s)$ for arbitrary $s \not\equiv 0 \pmod{p}$. Lemma 5.3 now follows from formula (5.1). $\square$

Now consider the mapping $\rho : \mathfrak{O} \to \mathbb{Z}/p\mathbb{Z}$ sending $\alpha = a_0 + a_1\zeta + \ldots + a_{p-2}\zeta^{p-2}$, $a_i \in \mathbb{Z}$, to $\rho(\alpha) \equiv a_0 + a_1 + \ldots + a_{p-2} \pmod{p}$. This is an additive homomorphism. Using Lemma 5.1, the kernel of this homomorphism is easily seen to equal $\mathfrak{P}$. This shows that

$$\mathfrak{O}/\mathfrak{P} \cong \mathbb{Z}/p\mathbb{Z} = \mathbb{F}_p,$$

and the mapping $\rho$ can be considered as the reduction mod $\mathfrak{P}$.

By the general remarks above, $\mathfrak{O}$ is isomorphic to $\mathbb{Z}^{p-1}$ as a $\mathbb{Z}$-module. We shall define a bilinear form on $\mathfrak{O}$.

Consider again the general case of an algebraic number field $K$ of degree $n = [K : \mathbb{Q}]$. Let $\mathfrak{O}$ be the ring of integers of $K$. Let $\sigma_1, \ldots, \sigma_n : K \longrightarrow \mathbb{C}$ be the different embeddings of $K$ in $\mathbb{C}$, and let $(\alpha_1, \ldots, \alpha_n)$ be a basis of the $\mathbb{Z}$-module $\mathfrak{O}$. Then $(\sigma_i(\alpha_j))$ is an $n \times n$-matrix. The number

$$\Delta_K := (\det(\sigma_i(\alpha_j)))^2$$

is independent of the choice of the $\sigma_i$ and the basis $(\alpha_1, \ldots, \alpha_n)$ of $\mathfrak{O}$, and is called the *discriminant* of $K$.

Now for $x, y \in \mathfrak{O}$ the mapping $(x, y) \mapsto \mathrm{Tr}(xy)$ defines a *symmetric bilinear form* on the $\mathbb{Z}$-module $\mathfrak{O}$ with values in $\mathbb{Z}$. So $(\mathrm{Tr}(\alpha_i\alpha_j))$ is an integral $n \times n$-matrix. Then

$$\begin{aligned}
\det(\mathrm{Tr}(\alpha_i\alpha_j)) &= \det\left(\sum_k \sigma_k(\alpha_i\alpha_j)\right) \\
&= \det\left(\sum_k \sigma_k(\alpha_i)\sigma_k(\alpha_j)\right) \\
&= \det(\sigma_k(\alpha_i)) \cdot \det(\sigma_k(\alpha_i)) \\
&= \Delta_K.
\end{aligned}$$

Now assume that $K = \mathbb{Q}(\alpha)$, and let $f(x) = a_0 + a_1x + \ldots + a_nx^n$ be the minimal polynomial of $\alpha$ over $\mathbb{Q}$. Assume in addition that $a_i \in \mathbb{Z}$ for $i = 0, \ldots, n$ and that $a_n = 1$. Finally assume that $1, \alpha, \ldots, \alpha^{n-1}$ is a basis of the $\mathbb{Z}$-module $\mathfrak{O}$.

Then

$$\begin{aligned}
\Delta_K &= \left(\det\left(\sigma_i\left(\alpha^{j-1}\right)\right)\right)^2 \\
&= \det\begin{pmatrix} 1 & \sigma_1(\alpha) & \cdots & \sigma_1(\alpha)^{n-1} \\ 1 & \sigma_2(\alpha) & \cdots & \sigma_2(\alpha)^{n-1} \\ \vdots & \vdots & \ddots & \vdots \\ 1 & \sigma_n(\alpha) & \cdots & \sigma_n(\alpha)^{n-1} \end{pmatrix}^2 \\
&= \left(\prod_{i<j}(\sigma_i(\alpha) - \sigma_j(\alpha))\right)^2 \qquad \text{(Vandermonde)} \\
&= (-1)^{n(n-1)/2} \prod_{i \neq j}(\sigma_i(\alpha) - \sigma_j(\alpha)).
\end{aligned}$$

This shows that $\Delta_K$ is the discriminant of the minimal polynomial $f(x)$ of $\alpha$, since the $\sigma_i(\alpha)$, $i = 1, \ldots, n$, are the conjugates of $\alpha$, i.e., the roots of the minimal polynomial $f(x)$ of $\alpha$.

Now we specialize again to the case $\alpha = \zeta$, $K = \mathbb{Q}(\zeta)$. In this case we get

$$\begin{aligned}
\Delta_{\mathbb{Q}(\zeta)} &= (-1)^{\frac{p-1}{2}} \prod_{i \neq j} (\zeta^i - \zeta^j) \\
&= (-1)^{\frac{p-1}{2}} \prod_{i=1}^{p-1} \prod_{\substack{k=1 \\ k \not\equiv -i(p)}}^{p-1} \left(\zeta^i - \zeta^{i+k}\right) \\
&= (-1)^{\frac{p-1}{2}} \prod_{i=1}^{p-1} \zeta^i \prod_{\substack{k=1 \\ k \not\equiv -i(p)}}^{p-1} \left(1 - \zeta^k\right) \\
&= (-1)^{\frac{p-1}{2}} \left( \prod_{k=1}^{p-1} \left(1 - \zeta^k\right) \right)^{p-2} \zeta^{\frac{1}{2}p(p-1)} \\
&= (-1)^{\frac{p-1}{2}} p^{p-2} \qquad \text{(by formula (5.1))}.
\end{aligned}$$

The absolute value of the discriminant $\Delta_{\mathbb{Q}(\zeta)}$ also follows from Lemma 5.3.

For $x \in \mathbb{Q}(\zeta)$ let $\bar{x}$ denote the complex conjugate. We now consider the bilinear form $(x,y) \mapsto \operatorname{Tr}(x\bar{y})$, $x, y \in \mathfrak{O}$, on $\mathfrak{O}$. Since

$$\operatorname{Tr}(x\bar{x}) = \sum_{\sigma} \sigma(x)\overline{\sigma(x)} > 0$$

for $x \neq 0$, $x \in \mathfrak{O}$, where $\sigma$ runs through the embeddings of $\mathbb{Q}(\zeta)$, this form is positive definite. This form has determinant $p^{p-2}$, as is easily deduced from $\det(\operatorname{Tr}(\zeta^i\zeta^j)) = (-1)^{\frac{p-1}{2}} p^{p-2}$ and the fact that the mapping $x \to \bar{x}$, considered as an endomorphism of the $\mathbb{Q}$-vector space $\mathbb{Q}(\zeta)$, has determinant $(-1)^{\frac{p-1}{2}}$. Thus the symmetric bilinear form

$$\langle x, y \rangle := \operatorname{Tr}\left(\frac{x\bar{y}}{p}\right), \qquad x, y \in \mathfrak{O},$$

has determinant $\frac{1}{p}$.

We have seen above that $\mathfrak{P} = (1 - \zeta)$ is a submodule of $\mathfrak{O}$ of index $p$. We can consider the bilinear form $\langle x, y \rangle = \operatorname{Tr}\left(\frac{x\bar{y}}{p}\right)$ on $\mathfrak{P}$: By the foregoing it has determinant $p$, and $\operatorname{Tr}\left(\frac{x\bar{y}}{p}\right) \in \mathbb{Z}$ for $x, y \in \mathfrak{P}$, by Lemma 5.2. Moreover this form is even: $\langle x, x \rangle = \operatorname{Tr}\left(\frac{x\bar{x}}{p}\right) \in 2\mathbb{Z}$ for all $x \in \mathfrak{P}$. For this we use the following remark.

**Remark 5.1** The real subfield of $K$ is

$$K \cap \mathbb{R} = \{x \in K \mid x = \bar{x}\}.$$

We denote this subfield by $k$. Let $\alpha \in K$, $\alpha = a_0 + a_1\zeta + \ldots + a_{p-1}\zeta^{p-1}$ with $a_i \in \mathbb{Q}$. Then $\alpha \in k$ if and only if $a_i = a_{p-i}$ for $i = 1, \ldots, \frac{p-1}{2}$. Let $\alpha \in k$. Then $\alpha$ can be written

$$\alpha = a_0 + a_1(\zeta + \zeta^{-1}) + a_2(\zeta^2 + \zeta^{-2}) + \ldots + a_{\frac{p-1}{2}} \left(\zeta^{\frac{p-1}{2}} + \zeta^{-\frac{p-1}{2}}\right).$$

This shows that $k = \mathbb{Q}(\zeta + \zeta^{-1})$, and that $(1, \zeta + \zeta^{-1}, \zeta^2 + \zeta^{-2}, \ldots, \zeta^{\frac{p-1}{2}-1} + \zeta^{-\frac{p-1}{2}+1})$ is a basis of $k$ over $\mathbb{Q}$. In particular $[K : k] = 2$, $[k : \mathbb{Q}] = \frac{p-1}{2}$. Moreover

$$\mathrm{Tr}_{K/\mathbb{Q}}\left(\frac{x\bar{x}}{p}\right) = 2\,\mathrm{Tr}_{k/\mathbb{Q}}\left(\frac{x\bar{x}}{p}\right).$$

This shows that the above bilinear form on $\mathfrak{P}$ is even.

Recall from Sect. 1.1 that a (positive definite) integral lattice is a free $\mathbb{Z}$-module (free abelian group) $\Gamma$ of finite rank $n$ together with a positive definite symmetric bilinear form $\langle\ ,\ \rangle : \Gamma \times \Gamma \to \mathbb{Z}$ with values in $\mathbb{Z}$. This lattice is called *even*, if $\langle x, x\rangle \in 2\mathbb{Z}$ for all $x \in \Gamma$. So $\mathfrak{P}$ with the bilinear form $\langle x, y\rangle = \mathrm{Tr}\left(\frac{x\bar{y}}{p}\right)$ is an even integral lattice. In fact we can prove the following lemma:

**Lemma 5.4.** *The lattice $\mathfrak{P}$ with the bilinear form $(x,y) \mapsto \langle x, y\rangle = \mathrm{Tr}\left(\frac{x\bar{y}}{p}\right)$ is isomorphic to the root lattice $A_{p-1}$.*

*Proof.* By the definition of $\langle\ ,\ \rangle$ we have

$$\begin{aligned}
\langle 1, 1\rangle &= \frac{p-1}{p},\\
\langle \zeta^r, \zeta^r\rangle &= \mathrm{Tr}\left(\frac{\zeta^r\zeta^{-r}}{p}\right) = \frac{p-1}{p},\\
\langle \zeta^r, \zeta^s\rangle &= \mathrm{Tr}\left(\frac{\zeta^r\zeta^{-s}}{p}\right) = \mathrm{Tr}\left(\frac{\zeta^{r-s}}{p}\right)\\
&= \frac{1}{p}(\zeta + \zeta^2 + \ldots + \zeta^{p-1}) = -\frac{1}{p}, \qquad r \not\equiv s \pmod{p}.
\end{aligned}$$

This implies that

$$\langle \zeta^r - \zeta^s, \zeta^r - \zeta^s\rangle = \frac{p-1}{p} + \frac{p-1}{p} - 2\left(-\frac{1}{p}\right) = 2,$$

so the vectors $\zeta^r - \zeta^s$ for $r \not\equiv s \pmod{p}$ are roots in $\mathfrak{P}$. There are $p(p-1)$ vectors of this form. This number agrees with the number of roots of $A_{p-1}$. Now the vectors

$$1 - \zeta,\ \zeta - \zeta^2,\ \zeta^2 - \zeta^3, \ldots, \zeta^{p-2} - \zeta^{p-1}$$

form a basis of $\mathfrak{P}$, and the corresponding Coxeter-Dynkin diagram is the diagram

$$\underset{1-\zeta}{\bullet} \!\!-\!\!\!-\!\!\!-\!\! \underset{\zeta-\zeta^2}{\bullet} \!\!-\!\!\!-\!\!\!-\!\! \underset{\zeta^2-\zeta^3}{\bullet} \!\!-\!\!\!-\ \cdots\ -\!\!\!-\!\! \underset{\zeta^{p-2}-\zeta^{p-1}}{\bullet}$$

which is the Coxeter-Dynkin diagram of $A_{p-1}$. This proves Lemma 5.4. □

**Remark 5.2** Note that the dual lattice $\mathfrak{P}^*$ of $\mathfrak{P}$ is $\mathfrak{O}$: The inclusion $\mathfrak{O} \subset \mathfrak{P}^*$ follows from Lemma 5.2, which implies that $\mathrm{Tr}\left(\frac{x\bar{y}}{p}\right) \in \mathbb{Z}$ for $y \in \mathfrak{O}$, $x \in \mathfrak{P}$. In Sect. 1.1 we have shown that the index $[\Gamma^* : \Gamma]$ of a lattice $\Gamma$ in its dual lattice $\Gamma^*$ is equal to the discriminant $\mathrm{disc}\,(\Gamma)$ of $\Gamma$. So $[\mathfrak{P}^* : \mathfrak{P}] = p$. Since also $[\mathfrak{O} : \mathfrak{P}] = p$, we conclude that $\mathfrak{P}^* = \mathfrak{O}$.

## 5.2 Construction of Lattices from Codes over $\mathbb{F}_p$

As before, let $p$ be an odd prime number. Let $C \subset \mathbb{F}_p^n$ be a (linear) code over $\mathbb{F}_p$ of length $n$. We assume throughout this section that $C \subset C^{\perp}$. This means that $x \cdot y = 0$ for all $x, y \in C$. Let $m = \dim C$.

As in Sect. 5.1, let $\mathfrak{O}$ be the ring of integers of the cyclotomic field $\mathbb{Q}(\zeta)$. We construct a lattice $\Gamma_C \subset \mathfrak{O}^n$ from $C$ as follows. By Sect. 5.1, $\Gamma_C$ will be an integral lattice of rank $n(p-1)$. The construction generalizes the construction of Sect. 1.3 for the case $p = 2$.

**Definition.** Let $\rho : \mathfrak{O}^n \to \mathbb{F}_p^n$ be the mapping defined by the reduction modulo the principal ideal $\mathfrak{P} = (1 - \zeta)$ in each coordinate. Then define

$$\Gamma_C := \rho^{-1}(C) \subset \mathfrak{O}^n.$$

Let $x, y \in \mathfrak{O}^n$, $x = (x_1, \ldots, x_n)$, $y = (y_1, \ldots, y_n)$ with $x_i, y_i \in \mathfrak{O}$ for $i = 1, \ldots, n$. For each coordinate we have the symmetric bilinear form defined by $(x_i, y_i) \mapsto \mathrm{Tr}\left(\frac{x_i \overline{y_i}}{p}\right)$. We define a symmetric bilinear form on $\mathfrak{O}^n$ by

$$\langle x, y \rangle = \sum_{i=1}^{n} \mathrm{Tr}\left(\frac{x_i \overline{y_i}}{p}\right).$$

Let us define $x \cdot \bar{y} = x_1 \overline{y_1} + \ldots + x_n \overline{y_n}$, $x \cdot y = x_1 y_1 + \ldots + x_n y_n$. Then

$$x \cdot \bar{y} \equiv x \cdot y \pmod{\mathfrak{P}},$$

since, for any $\alpha \in \mathfrak{O}$, one has $\alpha \equiv \overline{\alpha} \pmod{\mathfrak{P}}$ (for proving this it suffices to consider the case $\alpha = \zeta^i$). Now assume that $\rho(x), \rho(y) \in C$. Then

$$0 \equiv x \cdot y \equiv x \cdot \bar{y} \pmod{\mathfrak{P}}.$$

By Sect. 5.1, this implies that

$$\langle x, y \rangle \in \mathbb{Z},$$
$$\langle x, x \rangle \in 2\mathbb{Z},$$

for all $x, y \in \Gamma_C \subset \mathfrak{O}^n$. Therefore the lattice $\Gamma_C$ is an even integral lattice of rank $n(p-1)$.

By Sect. 5.1 the bilinear form $\langle\ ,\ \rangle$ on $\mathfrak{O}^n$ has determinant $\left(\frac{1}{p}\right)^n$. Since $\rho$ is surjective, $\rho^{-1}(C)$ has index $p^{n-m}$ in $\mathfrak{O}^n$, where $m = \dim C$. Therefore the discriminant of the lattice

$\Gamma_C$ is

$$\mathrm{disc}\,(\Gamma_C) = p^{n-2m}.$$

**Lemma 5.5.** *Let $C \subset \mathbb{F}_p^n$ be a linear code of dimension $m$ with $C \subset C^\perp$. Then*

$$\Gamma_C^* = \Gamma_{C^\perp}.$$

*Proof.* Let $x \in \Gamma_C$, $y \in \Gamma_{C^\perp}$. Then $\rho(x) \in C$, $\rho(y) \in C^\perp$, and we have $\rho(x) \cdot \rho(y) \equiv 0 \pmod{p}$. Therefore $x \cdot y \in \mathfrak{P}$, and hence $\langle x, y\rangle \in \mathbb{Z}$. This proves $\Gamma_{C^\perp} \subset \Gamma_C^*$.

Now

$$\mu = \mathrm{vol}\,(\mathbb{R}^{n(p-1)}/\Gamma_C) = p^{\frac{n}{2}-m}.$$

Since $C^\perp$ has dimension $n-m$, we have

$$\mathrm{disc}\,(\Gamma_{C^\perp}) = p^{n-2(n-m)} = p^{2m-n},$$

and hence

$$\mathrm{vol}\,(\mathbb{R}^{n(p-1)}/\Gamma_{C^\perp}) = p^{m-\frac{n}{2}} = \frac{1}{\mu}.$$

This implies that $\Gamma_{C^\perp} = \Gamma_C^*$. This proves Lemma 5.5. □

Putting everything together we get:

**Proposition 5.2.** *Let $C \subset \mathbb{F}_p^n$ be a linear code of dimension $m$ with $C \subset C^\perp$. Then the lattice $\Gamma_C$, provided with the symmetric bilinear form $\langle x,y\rangle = \sum_{i=1}^{n} \mathrm{Tr}\left(\frac{x_i\overline{y_i}}{p}\right)$, is an even integral lattice of rank $n(p-1)$ and discriminant $p^{n-2m}$. If $C$ is self-dual, then $\Gamma_C$ is unimodular.*

**Corollary 5.1.** *If $C$ is a self-dual code over $\mathbb{F}_p$, then $n(p-1) \equiv 0 \pmod{8}$.*

*Proof.* This follows from Proposition 5.2 and Theorem 2.1. □

Note that the lattice $\Gamma_C$ contains the root lattice $nA_{p-1}$, since $\mathfrak{P}^n \subset \Gamma_C$.

**Remark 5.3** Let $\Gamma \subset \mathbb{R}^l$ be an irreducible root lattice. Then the root lattices $\Lambda$ of maximal dimension $l$ contained in $\Gamma$ can be obtained by the following algorithm, due to A. Borel and J. de Siebenthal [3]. Consider the extended Coxeter-Dynkin diagram corresponding to $\Gamma$ (cf. Sect. 1.5). Remove one vertex from this diagram. Then the remaining graph describes a root lattice of dimension $l$ contained in $\Gamma$, and all such root lattices are obtained in this way by repeating this process an appropriate number of times. One gets Table 5.1 of inclusions of root lattices which is adopted from the paper of Borel and de Siebenthal [3]. Here $D_2 = A_1 \perp A_1$, $D_3 = A_3$. Note that we have already considered a special case of these inclusions, namely the case $lA_1 \subset \Gamma$ in Proposition 1.5. Concerning the inclusions $nA_{p-1} \subset \Gamma$, $n(p-1) = l$, $p$ an odd prime number, we get from this table the only inclusions

$$3A_2 \subset E_6,$$
$$4A_2 \subset E_8,$$
$$2A_4 \subset E_8.$$

**Table 5.1** Maximal root sublattices of the irreducible root lattices

| $\Gamma$ | maximal root sublattice |
|---|---|
| $A_l$ | — |
| $D_l$ | $D_i \perp D_{l-i} \quad (i=2,3,\ldots,l-2)$ |
| $E_6$ | $A_1 \perp A_5,\ A_2 \perp A_2 \perp A_2$ |
| $E_7$ | $A_1 \perp D_6,\ A_7,\ A_2 \perp A_5$ |
| $E_8$ | $D_8,\ A_1 \perp E_7,\ A_8,\ A_2 \perp E_6,\ A_4 \perp A_4$ |

We may try to find a linear code $C \subset \mathbb{F}_p^n$ with $\Gamma_C = \Gamma$ in each of these cases.

**Example 5.1** Consider the ternary linear code

$$C = \{(0,0,0), (1,1,1), (-1,-1,-1)\} \subset \mathbb{F}_3^3.$$

This is a one-dimensional code with $C \subset C^\perp$. Therefore the lattice $\Gamma_C$ is defined; it is an even lattice of rank 6 and discriminant 3. Since it contains the root lattice $3A_2$, but also roots not contained in this sublattice, e.g. $(1,1,1) \in \mathfrak{O}^3$, it contains the root lattice $E_6$ by the above remark. Since this is a lattice of the same rank and discriminant, $\Gamma_C$ is equal to $E_6$.

**Example 5.2** Consider the linear code $C \subset \mathbb{F}_3^4$ with the generator matrix

$$\begin{pmatrix} 1\,1\,1\,0 \\ 1\,2\,0\,1 \end{pmatrix}.$$

This code is self-dual. Therefore the lattice $\Gamma_C$ is an even unimodular lattice of rank 8, thus equal to $E_8$.

**Example 5.3** Consider the linear code

$$C = \{(0,0), (1,2), (2,-1), (-2,1), (-1,-2)\} \subset \mathbb{F}_5^2.$$

This is also a self-dual code. Again the lattice $\Gamma_C$ is an even unimodular lattice of rank 8, thus $\Gamma_C = E_8$.

**Exercise 5.1** In each of the above examples, determine the roots in $\Gamma_C$.

For the dimension $24 = n(p-1)$ the prime numbers $p = 3, 5, 7, 13$ are interesting. For $p = 3$ consider the linear code $C_1 \subset \mathbb{F}_3^{12}$ defined by the generator matrix

$$\left(\begin{array}{cccccc|cccccc} 1&0&0&0&0&0&0&1&1&1&1&1 \\ 0&1&0&0&0&0&1&0&1&2&2&1 \\ 0&0&1&0&0&0&1&1&0&1&2&2 \\ 0&0&0&1&0&0&1&2&1&0&1&2 \\ 0&0&0&0&1&0&1&2&2&1&0&1 \\ 0&0&0&0&0&1&1&1&2&2&1&0 \end{array}\right).$$

This is a $[12,6]$-code. One can show that $C_1 \subset C_1^{\perp}$. One can either check this directly (tedious) or apply [56, (1.3.8)]. Therefore $C_1$ is a self-dual code which is called the *extended ternary Golay code*. Its minimum weight is 6. Therefore $\Gamma_{C_1}$ is an even unimodular lattice of rank 24 containing the root lattice $12A_2$ but no other roots.

The code $C_1$ has 132 pairs of codewords of weight 6 equivalent under multiplication by $-1$. These are 132 6-subsets of a 12-set. They form a Steiner system $S(5,6,12)$ which can be shown to be unique up to permutations. From this it follows that the extended ternary Golay code is unique up to equivalence under the symmetric group. The automorphism group of the Steiner system $S(5,6,12)$ is the Mathieu group $M_{12}$. Therefore the automorphism group of the extended ternary Golay code is the semi-direct product of $M_{12}$ and the group $\mathbb{Z}_2$, acting by multiplication by $-1$ (see [66, Chap. 20]).

Now consider the code $C_2 \subset \mathbb{F}_3^{12}$ defined by the generator matrix

$$\begin{pmatrix} 111\ 000\ 000\ 000 \\ 000\ 111\ 000\ 000 \\ 000\ 000\ 111\ 000 \\ 000\ 000\ 000\ 111 \\ 120\ 120\ 120\ 000 \\ 000\ 021\ 012\ 012 \end{pmatrix}.$$

This is a $[12,6]$-code, too. One easily checks that $C_2 \subset C_2^{\perp}$, hence $C_2$ is self-dual. By Example 5.1 the lattice $\Gamma_{C_2}$ contains the root lattice $4E_6$. Therefore $\Gamma_{C_2}$ is an even unimodular lattice of rank 24 containing $4E_6$.

For $p=5$ let $C \subset \mathbb{F}_5^6$ be the linear code defined by the generator matrix

$$\begin{pmatrix} 0 & 0 & 1 & 1 & 2 & -2 \\ 2 & -2 & 0 & 0 & 1 & 1 \\ 1 & 1 & 2 & -2 & 0 & 0 \end{pmatrix}.$$

Then the lattice $\Gamma_C$ is an even unimodular lattice of rank 24 containing $6A_4$.

For $p=7$ we consider the linear code $C \subset \mathbb{F}_7^4$ with the generator matrix

$$\begin{pmatrix} 1 & 2 & 3 & 0 \\ 0 & 3 & -2 & 1 \end{pmatrix}.$$

The lattice $\Gamma_C$ is an even unimodular lattice of rank 24 containing $4A_6$.

For $p=13$ let $C \subset \mathbb{F}_{13}^2$ be the linear code generated by the vector $(1,5)$. The lattice $\Gamma_C$ is an even unimodular lattice of rank 24 containing $2A_{12}$.

Summarizing we have shown the existence of the 5 Niemeier lattices with root lattices

$$12A_2,\ 4E_6,\ 6A_4,\ 4A_6,\ 2A_{12}.$$

One can show that all the self-dual codes needed to construct these lattices are unique up to equivalence. By Corollary 3.6 the corresponding lattices are therefore unique up to isomorphism.

Combining these results with Corollary 3.7 we have proved Theorem 3.4 for 14 of the 24 cases.

## 5.3 Theta Functions over Number Fields

We shall associate a theta function to the lattice $\Gamma_C$. This will be a function from the product of $\frac{p-1}{2}$ upper half planes $\mathbb{H}$ to $\mathbb{C}$, as studied, for example, by H. D. Kloosterman [43] and M. Eichler [24].

Let us first consider the case $\Gamma = A_{p-1} \subset \mathbb{R}^{p-1}$, where $p$ is an odd prime number as before. Then $\Gamma^*/\Gamma = \mathbb{Z}/p\mathbb{Z}$. Let $\rho \in \Gamma^*$. In Sect. 3.1 we have considered the theta function

$$\vartheta_{\rho+\Gamma}(z) := \sum_{x \in \rho+\Gamma} e^{\pi i z x^2}, \qquad z \in \mathbb{H}.$$

It was shown that $\vartheta_{\rho+\Gamma}$ depends only on $\rho \in \Gamma^*/\Gamma$ and does not change under the substitution $\rho \mapsto -\rho$. The lattice $A_{p-1}$ has level $N = p$. So Theorem 3.2 yields in this case

$$\vartheta_{\rho+\Gamma}\left(\tfrac{az+b}{cz+d}\right) = (cz+d)^{\frac{p-1}{2}} \vartheta_{\rho+\Gamma}(z) \quad \text{for} \begin{pmatrix} a & b \\ c & d \end{pmatrix} \in \Gamma(p),$$

$$\vartheta_{\Gamma}\left(\tfrac{az+b}{cz+d}\right) = (cz+d)^{\frac{p-1}{2}} \left(\tfrac{d}{p}\right) \vartheta_{\Gamma}(z) \text{ for} \begin{pmatrix} a & b \\ c & d \end{pmatrix} \in \Gamma_0(p),$$

where for the second identity we use

$$\left(\frac{\Delta}{d}\right) = \left(\frac{(-1)^{(p-1)/2} p}{d}\right) = \left(\frac{d}{p}\right)$$

by the quadratic reciprocity law (cf. [81]). Since $\Gamma^*/\Gamma = \mathbb{Z}/p\mathbb{Z}$, we have to consider $\rho = 0, 1, \ldots, \frac{p-1}{2}$. Hence we get $\frac{p-1}{2} + 1$ theta functions

$$\vartheta_j := \vartheta_{j+\Gamma}, \qquad \text{for } j = 0, 1, \ldots, \frac{p-1}{2}.$$

For our purposes it is not sufficient to consider these theta functions of one variable. We shall now introduce appropriate generalized theta functions of several variables. Consider again the field $K = \mathbb{Q}(\zeta)$ where $\zeta = e^{2\pi i/p}$. Let $k = \mathbb{Q}(\zeta + \zeta^{-1})$ be the real subfield. Let $\mathfrak{O}$ be the ring of integers in $K$, and let $\mathfrak{P}$ be the principal ideal of $\mathfrak{O}$ generated by the element $1-\zeta$. By Lemma 5.4 we can identify the lattice $\Gamma$ with the lattice $\mathfrak{P}$ provided with the symmetric bilinear form $\langle x, y \rangle = \mathrm{Tr}_{K/\mathbb{Q}}\left(\frac{x\bar{y}}{p}\right)$, and the dual lattice $\Gamma^*$ with $\mathfrak{O}$. Then

$$\vartheta_j(z) = \sum_{x \in \mathfrak{P}+j} e^{\pi i z \mathrm{Tr}_{K/\mathbb{Q}}\left(\frac{x\bar{x}}{p}\right)}$$

$$= \sum_{x \in \mathfrak{P}+j} e^{2\pi i z \mathrm{Tr}_{k/\mathbb{Q}}\left(\frac{x\bar{x}}{p}\right)}$$

for $j = 0, 1, \ldots, \frac{p-1}{2}$, by Remark 5.1.

Since $k$ is the real subfield of $K$ and $[k : \mathbb{Q}] = \frac{p-1}{2}$, there exist exactly $\frac{p-1}{2}$ distinct real embeddings $\sigma_l : k \to \mathbb{R}$, $l = 1, \ldots, \frac{p-1}{2}$. Each of these $\sigma_l$ is of the form $\zeta + \zeta^{-1} \mapsto \zeta^a + \zeta^{-a}$ for a suitable integer $a$. In particular, one has $\sigma_l(k) = k$. Hence the $\sigma_l$ form a group (with respect to composition), the Galois group of $k$ over $\mathbb{Q}$. We denote it by $\mathrm{Gal}(k, \mathbb{Q})$. Consider the product

$$\mathbb{H}^{\frac{p-1}{2}} = \mathbb{H} \times \ldots \times \mathbb{H} \qquad ((p-1)/2 \text{ times})$$

of $\frac{p-1}{2}$ upper half planes. Let $z = (z_1, \ldots, z_{(p-1)/2})$ be a point of $\mathbb{H}^{(p-1)/2}$. We define the theta function $\theta_j(z)$ depending on $\frac{p-1}{2}$ variables $z_l \in \mathbb{H}$ by

$$\theta_j(z) := \sum_{x \in \mathfrak{P}+j} e^{2\pi i \mathrm{Tr}_{k/\mathbb{Q}}\left(z\frac{x\bar{x}}{p}\right)},$$

where

$$\mathrm{Tr}_{k/\mathbb{Q}}\left(z\frac{x\bar{x}}{p}\right) := \sum_{l=1}^{(p-1)/2} z_l \cdot \frac{\sigma_l(x\bar{x})}{p}.$$

Setting all variables $z_l$ equal to one variable $z_m$ yields the old function $\vartheta_j$. The function $\theta_j$ is holomorphic in $z \in \mathbb{H}^{(p-1)/2}$.

Let $\mathfrak{O}_k$ be the ring of integers in $k$. The group $\mathrm{SL}_2(\mathfrak{O}_k)$ is the group of all $2 \times 2$-matrices

$$\begin{pmatrix} \alpha & \beta \\ \gamma & \delta \end{pmatrix}$$

with entries $\alpha, \beta, \gamma, \delta \in \mathfrak{O}_k$ and with determinant $\alpha\delta - \beta\gamma = 1$. This group operates on $\mathbb{H}^{(p-1)/2}$ by

$$z \longmapsto \frac{\alpha z + \beta}{\gamma z + \delta}, \qquad z_l \longmapsto \frac{\sigma_l(\alpha) z_l + \sigma_l(\beta)}{\sigma_l(\gamma) z_l + \sigma_l(\delta)}, \qquad l = 1, \ldots, \frac{p-1}{2}.$$

The norm $N_{k/\mathbb{Q}}(\alpha)$ of an element $\alpha \in k$ over $\mathbb{Q}$ is defined by

$$N_{k/\mathbb{Q}}(\alpha) := \prod_{l=1}^{(p-1)/2} \sigma_l(\alpha).$$

For $z \in \mathbb{H}^{(p-1)/2}$, $\gamma, \delta \in \mathfrak{O}_k$, we define

$$N_{k/\mathbb{Q}}(\gamma z + \delta) := \prod_{l=1}^{(p-1)/2} \left(\sigma_l(\gamma) z_l + \sigma_l(\delta)\right).$$

For $\sigma \in \mathrm{Gal}(k,\mathbb{Q})$ we set $\sigma(z) = (z_{\varepsilon(1)},\ldots,z_{\varepsilon(\frac{p-1}{2})})$, where $\varepsilon$ denotes that permutation of the indices $1,\ldots,\frac{p-1}{2}$ such that $\sigma_l \circ \sigma = \sigma_{\varepsilon(l)}$ for $1 \leq l \leq \frac{p-1}{2}$. Finally let $\Gamma$ be a subgroup of $\mathrm{SL}_2(\mathfrak{O}_k)$.

**Definition.** A holomorphic function $f : \mathbb{H}^{(p-1)/2} \to \mathbb{C}$ is called a *Hilbert modular form of weight $m$ for $\Gamma$*, if

$$f\left(\frac{\alpha z+\beta}{\gamma z+\delta}\right) = f(z)\cdot N_{k/\mathbb{Q}}(\gamma z+\delta)^m \qquad \text{for all } \begin{pmatrix} \alpha & \beta \\ \gamma & \delta \end{pmatrix} \in \Gamma.$$

It is called *symmetric*, if $f(\sigma(z)) = f(z)$ for all $\sigma \in \mathrm{Gal}(k,\mathbb{Q})$.

The condition of holomorphicity at the cusps is superfluous for $p > 3$, see [26, p. 18]. For a recent account on Hilbert modular forms see also [25].

Let $\mathfrak{p}$ be the ideal $\mathfrak{p} := \mathfrak{P} \cap \mathfrak{O}_k$ of $\mathfrak{O}_k$ where $\mathfrak{O}_k$ denotes the ring of integers of $k$. Then

$$\mathfrak{p} = (\zeta+\zeta^{-1}-2) = \left((\zeta-1)(\zeta^{-1}-1)\right).$$

By Lemma 5.3 we have

$$\mathfrak{p}^{\frac{p-1}{2}} = (p).$$

Analogously to Sect. 3.1 we define

$$\Gamma(\mathfrak{p}) := \left\{ \begin{pmatrix} \alpha & \beta \\ \gamma & \delta \end{pmatrix} \in \mathrm{SL}_2(\mathfrak{O}_k) \;\middle|\; \begin{array}{l} \alpha \equiv \delta \equiv 1 \pmod{\mathfrak{p}}, \\ \beta \equiv \gamma \equiv 0 \pmod{\mathfrak{p}} \end{array} \right\},$$

$$\Gamma_0(\mathfrak{p}) := \left\{ \begin{pmatrix} \alpha & \beta \\ \gamma & \delta \end{pmatrix} \in \mathrm{SL}_2(\mathfrak{O}_k) \;\middle|\; \gamma \equiv 0 \pmod{\mathfrak{p}} \right\}.$$

Then we have the following result.

**Theorem 5.1.** *The function $\theta_j$, $j = 0,1,\ldots,\frac{p-1}{2}$, is a Hilbert modular form of weight 1 for the group $\Gamma(\mathfrak{p})$. Moreover one has*

$$\theta_0\left(\frac{\alpha z+\beta}{\gamma z+\delta}\right) = \theta_0(z)\cdot\left(\frac{\delta}{p}\right)\cdot N_{k/\mathbb{Q}}(\gamma z+\delta) \qquad \textit{for all } \begin{pmatrix} \alpha & \beta \\ \gamma & \delta \end{pmatrix} \in \Gamma_0(\mathfrak{p}).$$

Now let $C \subset \mathbb{F}_p^n$ be a self-dual code. By Corollary 5.1 we have $n(p-1) \equiv 0 \pmod 8$. Let $\Gamma_C$ be the lattice constructed from $C$ in Sect. 5.2. Then $\Gamma_C$ is an even unimodular lattice of rank $n(p-1)$. By the definition of the symmetric bilinear form on $\Gamma_C$, the usual theta function of the lattice $\Gamma_C$ is the function

$$\vartheta_C(z) = \sum_{x\in\Gamma_C} e^{2\pi i z \mathrm{Tr}_{k/\mathbb{Q}}\left(\frac{x\bar{x}}{p}\right)}, \qquad \text{where } z \in \mathbb{H}.$$

This is a modular form in one variable $z \in \mathbb{H}$. Now $\Gamma_C$ is not only a $\mathbb{Z}$-module, but also an $\mathfrak{O}_k$-module. As above we can define a theta function in several variables. For $z \in \mathbb{H}^{(p-1)/2}$ define

$$\theta_C(z) := \sum_{x \in \Gamma_C} e^{2\pi i \mathrm{Tr}_{k/\mathbb{Q}}\left(z\frac{x\bar{x}}{p}\right)}.$$

**Theorem 5.2.** *The function $\theta_C$ is a Hilbert modular form of weight $n$ for the whole group $\mathrm{SL}_2(\mathfrak{O}_k)$.*

The proofs of Theorem 5.1 and Theorem 5.2 will be given in Sect. 5.7.

**Remark 5.4** The Hilbert modular forms $\theta_j$ and $\theta_C$ are symmetric (in the sense of the definition above). This is due to the fact that the lattices $\mathfrak{P}$ and $\Gamma_C$ are invariant under the obvious action of the Galois group of $\mathbb{Q}(\zeta)$ over $\mathbb{Q}$.

The *Lee weight enumerator* of a code $C \subset \mathbb{F}_p^n$ is the polynomial

$$W_C\left(X_0, X_1, \ldots, X_{\frac{p-1}{2}}\right) := \sum_{u \in C} X_0^{l_0(u)} X_1^{l_1(u)} \ldots X_{\frac{p-1}{2}}^{l_{(p-1)/2}(u)},$$

where $l_0(u)$ is the number of zeros in $u$, and $l_i(u)$, for $i = 1, \ldots, \frac{p-1}{2}$, is the number of $+i$ or $-i$ occurring in the codeword $u$. This is a homogeneous polynomial of degree $n$.

We can now formulate the main theorem of G. van der Geer and F. Hirzebruch.

**Theorem 5.3 (van der Geer, Hirzebruch, Theorem of Alpbach[1]).** *Let $C \subset \mathbb{F}_p^n$ be a linear code with $C \subset C^\perp$. Then the following identity holds:*

$$\theta_C = W_C\left(\theta_0, \theta_1, \ldots, \theta_{\frac{p-1}{2}}\right).$$

*Proof.* Let $c$ be a codeword of $C$. Then

$$\sum_{x \in \rho^{-1}(c)} e^{2\pi i \mathrm{Tr}_{k/\mathbb{Q}}\left(z\frac{x\bar{x}}{p}\right)} = \theta_0^{l_0(c)}(z)\theta_1^{l_1(c)}(z) \ldots \theta_{\frac{p-1}{2}}^{l_{(p-1)/2}(c)}(z),$$

since

$$\theta_j(z) = \sum_{x \in \mathfrak{P}+j} e^{2\pi i \mathrm{Tr}_{k/\mathbb{Q}}\left(z\frac{x\bar{x}}{p}\right)} = \sum_{x \in \mathfrak{P}-j} e^{2\pi i \mathrm{Tr}_{k/\mathbb{Q}}\left(z\frac{x\bar{x}}{p}\right)}$$

and $\rho(\mathfrak{P} \pm j) = \pm j \in \mathbb{F}_p$. Summing over all codewords in $C$ yields Theorem 5.3. □

A similar relation between the complete weight enumerator of a linear code $C \subset \mathbb{F}_p^n$ with $C \subset C^\perp$ and certain Jacobi forms over the field $\mathbb{Q}(\zeta + \zeta^{-1})$ was recently obtained in [10].

We shall now give applications of the results of this section.

---

[1] Van der Geer and Hirzebruch like to call this theorem the Theorem of Alpbach because it was found during their summer school on coding theory in 1985 in Alpbach in the Austrian alps.

## 5.4 The Case $p = 3$: Ternary Codes

In this section we consider the case $p = 3$. In this case $C$ is a ternary linear code. The field $K$ is the field $\mathbb{Q}(\zeta)$, where

$$\zeta = e^{2\pi i/3} = -\frac{1}{2} + \frac{1}{2}\sqrt{-3}.$$

The real subfield $k = \mathbb{Q}(\zeta + \zeta^{-1})$ of $K$ is equal to $\mathbb{Q}$. The Hilbert modular forms are therefore usual modular forms. Theorem 5.3 reads as follows:

$$W_C(\theta_0, \theta_1) = \theta_C.$$

We consider the functions $\theta_0$ and $\theta_1$. The lattice $\mathfrak{P}$ is isomorphic to $A_2$ (Lemma 5.4). The lattice $A_2$ is the $\mathbb{Z}$-module $\mathbb{Z}^2$ with the quadratic form

$$(x, y)^2 = 2x^2 - 2xy + 2y^2.$$

Therefore

$$\begin{aligned} \theta_0(z) &= \sum_{(x,y)\in\mathbb{Z}^2} e^{\pi i z(2x^2 - 2xy + 2y^2)} \\ &= \sum_{(x,y)\in\mathbb{Z}^2} q^{x^2 - xy + y^2}, \end{aligned}$$

with $q = e^{2\pi i z}$. From this formula, one easily computes that the expansion of $\theta_0$ in $q$ starts as follows (cf. [21, p. 111]):

$$\theta_0(z) = 1 + 6\left(q + q^3 + q^4 + 2q^7 + q^9 + q^{12} + 2q^{13} + \ldots\right).$$

Similarly one can compute

$$\begin{aligned} \theta_1(z) &= q^{\frac{1}{3}} \sum_{(x,y)\in\mathbb{Z}^2} q^{x^2 - xy + y^2 + x - y} \\ &= 3q^{\frac{1}{3}}\left(1 + q + 2q^2 + 2q^4 + \ldots\right). \end{aligned}$$

The functions $\theta_0$ and $\theta_1$ are modular forms of weight 1 for the group

$$\Gamma(3) = \left\{ \begin{pmatrix} a & b \\ c & d \end{pmatrix} \in \mathrm{SL}_2(\mathbb{Z}) \;\middle|\; \begin{array}{l} a \equiv d \equiv 1 \pmod 3, \\ b \equiv c \equiv 0 \pmod 3 \end{array} \right\}$$

(Theorem 5.1). Then one has the following general result. Let $N$ be a positive integer, and let $\phi : \mathbb{Z} \to \mathbb{Z}/N\mathbb{Z}$ be the reduction modulo $N$. Then there is an induced homomorphism

$$\widetilde{\varphi} : \mathrm{SL}_2(\mathbb{Z}) \longrightarrow \mathrm{SL}_2(\mathbb{Z}/N\mathbb{Z}).$$

The principal congruence subgroup $\Gamma(N)$ of level $N$ is the kernel of this map and therefore a normal subgroup of $\mathrm{SL}_2(\mathbb{Z})$.

**Proposition 5.3.** *The mapping $\widetilde{\varphi}$ is surjective.*

*Proof.* Let $\begin{pmatrix} a & b \\ c & d \end{pmatrix}$, $ad-bc \equiv 1 \pmod N$, represent a matrix in $\mathrm{SL}_2(\mathbb{Z}/N\mathbb{Z})$. We can write the determinant condition in the form $ad-bc-mN=1$ for some integer $m$, hence $(c,d,N)=1$. We can therefore find an integer $k$ such that $(c, d+kN)=1$ (exercise!), and can thus assume that $(c,d)=1$. Therefore there exist integers $e$, $f$ such that $m=fc-ed$. Then the matrix

$$\begin{pmatrix} a+eN & b+fN \\ c & d \end{pmatrix}$$

is a matrix in $\mathrm{SL}_2(\mathbb{Z})$ representing the given matrix in $\mathrm{SL}_2(\mathbb{Z}/N\mathbb{Z})$. This proves Proposition 5.3. □

For $N=3$ the image of the mapping $\widetilde{\varphi}$ is the group $\mathrm{SL}_2(\mathbb{F}_3)$, so

$$\mathrm{SL}_2(\mathbb{F}_3) = \mathrm{SL}_2(\mathbb{Z})/\Gamma(3).$$

The group $\mathrm{PSL}_2(\mathbb{F}_3) \cong \mathrm{SL}_2(\mathbb{F}_3)/\{\pm 1\}$ is isomorphic to the alternating group $\mathscr{A}_4$, since $\mathrm{PSL}_2(\mathbb{F}_3)$ acts on the projective line over $\mathbb{F}_3$ preserving the cross ratio of the four points. The group $\mathscr{A}_4$ is also isomorphic to the *symmetry group of a tetrahedron*. This is a subgroup of $\mathrm{SO}_3(\mathbb{R})$ consisting of 12 elements.

Consider the group $\mathrm{SU}_2(\mathbb{C})$. This group is isomorphic to the three-dimensional sphere $S^3$, which can also be identified with the quaternions of absolute value 1. There is a mapping

$$S^3 \longrightarrow \mathrm{SO}_3(\mathbb{R})$$

sending a quaternion $q \in S^3$ to the rotation

$$x \longmapsto qxq^{-1}, \qquad x \in \mathbb{R}^3,$$

where $\mathbb{R}^3$ is identified with the purely imaginary quaternions. This mapping is a surjective homomorphism, its kernel is $\{\pm 1\}$, so

$$\mathrm{SO}_3(\mathbb{R}) = S^3/\{\pm 1\}.$$

The preimage $\widetilde{\mathscr{A}_4} \subset \mathrm{SU}_2(\mathbb{C}) \cong S^3$ of the symmetry group of the tetrahedron $\mathscr{A}_4 \subset \mathrm{SO}_3(\mathbb{R})$ is called the *binary tetrahedral group*; it has 24 elements. The group $\mathrm{SL}_2(\mathbb{F}_3)$ is isomorphic to the binary tetrahedral group $\widetilde{\mathscr{A}_4} \subset \mathrm{SU}_2(\mathbb{C})$.

The functions $\theta_0$ and $\theta_1$ are modular forms of weight 1 for $\Gamma(3)$. In fact they form a basis of all modular forms of weight 1 for $\Gamma(3)$. Moreover, the functions

$$\theta_0^k,\ \theta_0^{k-1}\theta_1, \ldots, \theta_0\theta_1^{k-1},\ \theta_1^k$$

form a basis of all modular forms of weight $k$ for $\Gamma(3)$. One has

**Theorem 5.4.** *The algebra of all modular forms for the group $\Gamma(3)$ is isomorphic to the polynomial algebra $\mathbb{C}[\theta_0, \theta_1]$.*

*Proof.* The dimension of the space of modular forms of weight $k$ for $\Gamma(3)$ can be derived by similar arguments as in Sect. 2.6, see [28, §8].

The Eisenstein series $E_4$ is the theta function of the unimodular $E_8$-lattice. We have seen in Example 5.2 that the lattice $E_8$ can be constructed as the lattice $\Gamma_C$ for a ternary self-dual code $C$ of length 4. This code $C$ has weight enumerator

$$X_0^4 + 8X_0X_1^3.$$

By Theorem 5.3 we get

$$E_4 = \theta_0^4 + 8\theta_0\theta_1^3.$$

This implies that $\theta_0$ and $\theta_1$ cannot be linearly dependent. For otherwise, $E_4$ would be a complex multiple of $\theta_1^4$. But this is impossible because $\theta_1^4$ is not a modular form, since

$$\theta_1^4 = q^{\frac{1}{3}} f(q),$$

where $f(q)$ is a power series in $q$.

We shall see below (see the proof of Proposition 5.4) that

$$\Delta = \frac{1}{1728}(E_4^3 - E_6^2) = -\frac{4^3}{1728}(\theta_1^4 - \theta_0^3\theta_1)^3.$$

Now the algebra of modular forms of weight divisible by 4 is the polynomial algebra $\mathbb{C}[E_4, \Delta]$. Therefore $\theta_0$ and $\theta_1$ are also algebraically independent. This proves Theorem 5.4. □

The group $\mathrm{SL}_2(\mathbb{Z})$ acts on the space of modular forms of weight 1 for $\Gamma(3)$ as follows. Let

$$\begin{pmatrix} a & b \\ c & d \end{pmatrix} \in \mathrm{SL}_2(\mathbb{Z}),$$

and let $\theta$ be a modular form of weight 1 for $\Gamma(3)$. Then

$$\theta\left(\frac{az+b}{cz+d}\right)(cz+d)^{-1}$$

is again a modular form of weight 1 for $\Gamma(3)$. The group $\Gamma(3)$ acts trivially. So $\mathrm{SL}_2(\mathbb{F}_3)$ acts on the space of modular forms of weight 1 for $\Gamma(3)$, hence, by Theorem 5.4, on $\mathbb{C}\cdot\theta_0 \oplus \mathbb{C}\cdot\theta_1$. It follows that $\mathrm{SL}_2(\mathbb{F}_3)$ acts on $\mathbb{C}[\theta_0, \theta_1]$ preserving the grading.

The generators $S$ and $T$ of $\mathrm{SL}_2(\mathbb{Z})$ act as follows. By definition of $\theta_0$ and $\theta_1$ and by Formula (T2) of Sect. 3.1, we have

$$\theta_0\left(-\frac{1}{z}\right) = \frac{z}{i\sqrt{3}}(\theta_0(z) + 2\theta_1(z)),$$
$$\theta_1\left(-\frac{1}{z}\right) = \frac{z}{i\sqrt{3}}(\theta_0(z) - \theta_1(z)).$$

Hence $S$ acts on $\mathbb{C}\cdot\theta_0+\mathbb{C}\cdot\theta_1$ by the linear mapping defined by

$$\frac{1}{i\sqrt{3}}\begin{pmatrix}1 & 2\\ 1 & -1\end{pmatrix}.$$

By Formula (T1) of Sect. 3.1, we have

$$\begin{aligned}\theta_0(z+1) &= \theta_0(z),\\ \theta_1(z+1) &= e^{\frac{2\pi i}{3}}\theta_1(z).\end{aligned}$$

Therefore $T$ acts on $\mathbb{C}\cdot\theta_0+\mathbb{C}\cdot\theta_1$ by the matrix

$$\begin{pmatrix}1 & 0\\ 0 & e^{\frac{2\pi i}{3}}\end{pmatrix}.$$

The modular group is the group $G=\mathrm{SL}_2(\mathbb{Z})/\{\pm 1\}$. It acts on the upper half plane $\mathbb{H}$ without fixed points. In Sect. 2.6 we have already remarked that one can compactify $\mathbb{H}/G$ by adding a point at $\infty$ to get a Riemann surface $\overline{\mathbb{H}/G}$ which is isomorphic to $\mathbb{P}_1(\mathbb{C})$. An isomorphism

$$j:\overline{\mathbb{H}/G}\xrightarrow{\cong}\mathbb{P}_1(\mathbb{C})$$

is given by the $j$-invariant:

$$j=\frac{E_4^3}{\Delta},\qquad \Delta=\frac{1}{1728}(E_4^3-E_6^2).$$

The mapping $j$ sends $\infty$ to $\infty$, and maps $\mathbb{H}/G$ bijectively onto $\mathbb{C}$.

Now consider $\Gamma(3)\subset\mathrm{SL}_2(\mathbb{Z})$. We have seen that $\mathrm{SL}_2(\mathbb{Z})/\Gamma(3)$ is the binary tetrahedral group. Since $-1\notin\Gamma(3)$, we may assume $\Gamma(3)\subset G$. The group $G/\Gamma(3)$ is the tetrahedral group. The theta functions $\theta_0$, $\theta_1$ define a mapping

$$\frac{\theta_0}{\theta_1}:\overline{\mathbb{H}/\Gamma(3)}\longrightarrow\mathbb{P}_1(\mathbb{C}).$$

This mapping is a bijection. The group $G/\Gamma(3)$ acts on $\overline{\mathbb{H}/\Gamma(3)}$. Under the above mapping this action corresponds to the action of the tetrahedral group on a tetrahedron lying inside the Riemann sphere $\mathbb{P}_1(\mathbb{C})$ as we shall see in a moment.

We have seen above that $\mathrm{SL}_2(\mathbb{F}_3)$ acts on the polynomial algebra $\mathbb{C}[\theta_0,\theta_1]$. The ring of invariant polynomials under this action is denoted by

$$\mathbb{C}[\theta_0,\theta_1]^{\mathrm{SL}_2(\mathbb{F}_3)}.$$

**Proposition 5.4.** $\mathbb{C}[\theta_0,\theta_1]^{\mathrm{SL}_2(\mathbb{F}_3)}=\mathbb{C}[E_4,E_6]$.

*Proof.* In the proof of Theorem 5.4, we have seen that

$$E_4=\theta_0^4+8\theta_0\theta_1^3.$$

This homogeneous polynomial in $\theta_0$ and $\theta_1$ has 4 zeros in $\mathbb{P}_1(\mathbb{C})$. Let us consider a sphere $S^2$ in $\mathbb{R}^3$ around the origin of radius $\sqrt{2}$ and the projection from the origin to the sphere $S^2$ of a tetrahedron which is inscribed in this sphere with the south pole as one vertex. We identify $S^2$ with $\mathbb{P}_1(\mathbb{C}) = \mathbb{C} \cup \{\infty\}$ under the stereographical projection from the north pole to the equatorial plane (cf. Fig. 5.1). The image of the tetrahedron under the stereo-

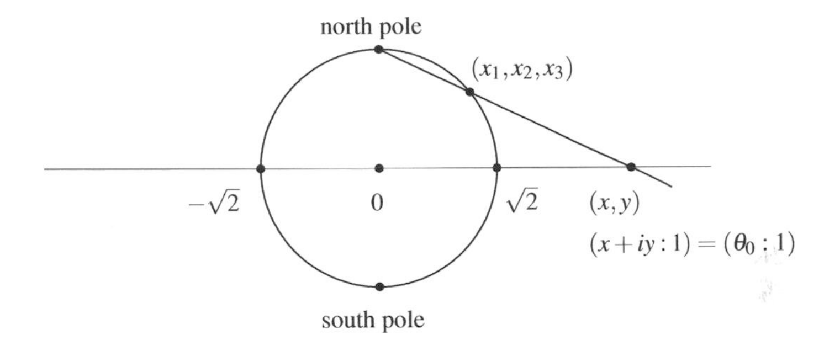


**Fig. 5.1** Stereographical projection of $S^2$ to $\mathbb{C} \cup \{\infty\}$

graphical projection is illustrated in Fig. 5.2. Let the homogeneous coordinates of $\mathbb{P}_1(\mathbb{C})$

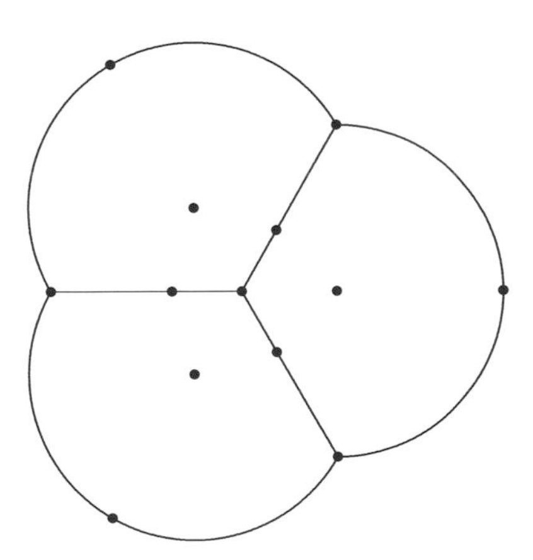

**Fig. 5.2** Stereographical projection of the tetrahedron

be $(\theta_0 : \theta_1)$ with $\infty = (1 : 0)$. Then the zeros of the homogeneous polynomial $\theta_0^4 + 8\theta_0\theta_1^3$ are just the vertices of this tetrahedron under this identification.

The action of $G/\Gamma(3)$ on $\mathbb{P}_1(\mathbb{C})$ is the action of the symmetry group of this tetrahedron.

Another system of 4 points invariant under this group are the midpoints of the faces of the tetrahedron. They are zeros of the homogeneous polynomial

$$\theta_1^4 - \theta_0^3\theta_1.$$

But this is not a modular form, since it is equal to $q^{1/3}$ times a power series in $q$. But

$$\left(\theta_1^4 - \theta_0^3\theta_1\right)^3$$

is a modular form of weight 12 for the whole group $G = \mathrm{SL}_2(\mathbb{Z})/\{\pm 1\}$. It is a cusp form, so it has to be a multiple of $\Delta$. By comparing coefficients, one can show that

$$\left(\theta_1^4 - \theta_0^3\theta_1\right)^3 = -\frac{1}{4^3}\left(E_4^3 - E_6^2\right).$$

In particular one can deduce from this formula and the formula which expresses $E_4$ in terms of $\theta_0$ and $\theta_1$ that

$$E_6 = \theta_0^6 - 20\theta_0^3\theta_1^3 - 8\theta_1^6.$$

The zeros of $E_6$ in $\mathbb{P}_1(\mathbb{C})$ correspond to the midpoints of the edges of the tetrahedron. This proves Proposition 5.4. □

**Theorem 5.5 (MacWilliams identity for ternary codes).** *Let $C \subset \mathbb{F}_3^n$ be a ternary linear code of dimension $m$ with $C \subset C^\perp$. Then*

$$W_{C^\perp}(X,Y) = \frac{1}{3^m} W_C(X+2Y, X-Y).$$

*Proof.* The assumption $C \subset C^\perp$ is unnecessary and the theorem can be proved within the framework of coding theory without this assumption, see [56, Theorem (3.5.3)]. However, as in the case of binary codes (Theorem 2.7), we again give a proof using modular forms. In order to be able to apply Theorem 5.3, we need the assumption $C \subset C^\perp$.

By Theorem 5.3 we have

$$\begin{aligned}
& W_C\left(\theta_0\left(-\frac{1}{z}\right), \theta_1\left(-\frac{1}{z}\right)\right) \\
&= \theta_C\left(-\frac{1}{z}\right) \\
&= \left(\frac{z}{i}\right)^n \frac{1}{\mathrm{vol}\,(\mathbb{R}^n/\Gamma_C)} \vartheta_{\Gamma_C^*}(z) \quad \text{(by Proposition 2.1)} \\
&= \left(\frac{z}{i}\right)^n \frac{1}{3^{\frac{n}{2}-m}} \vartheta_{\Gamma_{C^\perp}}(z) \quad \text{(by Lemma 5.5)} \\
&= \left(\frac{z}{i}\right)^n \frac{1}{3^{\frac{n}{2}-m}} W_{C^\perp}(\theta_0(z), \theta_1(z)).
\end{aligned}$$

On the other hand we have

$$W_C\left(\theta_0\left(-\frac{1}{z}\right), \theta_1\left(-\frac{1}{z}\right)\right) = \left(\frac{z}{i}\right)^n \frac{1}{3^{\frac{n}{2}}} W_C(\theta_0(z) + 2\theta_1(z), \theta_0(z) - \theta_1(z)).$$

Hence we get

$$W_{C^\perp}(\theta_0, \theta_1) = \frac{1}{3^m} W_C(\theta_0 + 2\theta_1, \theta_0 - \theta_1).$$

By Theorem 5.4, $\theta_0$ and $\theta_1$ are algebraically independent. This concludes the proof of Theorem 5.5. □

From Theorem 5.5 we immediately get the following corollary.

**Corollary 5.2.** *If $C \subset \mathbb{F}_3^n$ is a self-dual code, then the Hamming weight enumerator $W_C(X,Y)$ is invariant under the transformation*

$$\frac{1}{\sqrt{3}}\begin{pmatrix} 1 & 2 \\ 1 & -1 \end{pmatrix}$$

*in the $(X,Y)$-plane.*

We also get the following corollary proved by M. Broué and M. Enguehard [4].

**Corollary 5.3 (Broué, Enguehard).** *The Lee or Hamming weight enumerator of a self-dual ternary code is a polynomial in the modular forms $E_4$ and $E_6^2$.*

*Proof.* Let $W_C$ be the Lee or Hamming weight enumerator of a self-dual ternary code $C \subset \mathbb{F}_3^n$. By Corollary 5.1, $n$ is divisible by 4. We have to show that

$$W_C(\theta_0, \theta_1) \in \mathbb{C}[\theta_0, \theta_1]^{\mathrm{SL}_2(\mathbb{F}_3)}$$

Then the corollary follows from Proposition 5.4 and the fact that $n$ is divisible by 4.

We have seen above that the action of $S$ on $\mathbb{C}[\theta_0, \theta_1]$ is given by the matrix

$$\frac{1}{i\sqrt{3}}\begin{pmatrix} 1 & 2 \\ 1 & -1 \end{pmatrix}.$$

Using the fact that $n$ is divisible by 4 and Corollary 5.2, we see that $W_C(\theta_0, \theta_1)$ is invariant under $S$. The action of the generator $T$ on $\mathbb{C}[\theta_0, \theta_1]$ is given by the matrix

$$\begin{pmatrix} 1 & 0 \\ 0 & e^{\frac{2\pi i}{3}} \end{pmatrix}.$$

Since $C$ is self-dual, the weight of every codeword is divisible by 3. Therefore $\theta_1$ occurs in each monomial in $W_C(\theta_0, \theta_1)$ with a power divisible by 3. This shows that $W_C(\theta_0, \theta_1)$ is also invariant under $T$. We conclude that $W_C(\theta_0, \theta_1)$ is invariant under the action of $\mathrm{SL}_2(\mathbb{F}_3)$. This proves Corollary 5.3. □

**Example 5.4** Let us now consider the case $n = 12$. Let $C \subset \mathbb{F}_3^{12}$ be a self-dual code of length 12. By Proposition 5.4 and Corollary 5.3 the weight enumerator of $C$ has the form

$$\left(\theta_0^4+8\theta_0\theta_1^3\right)^3+a\left(\theta_1^4-\theta_0^3\theta_1\right)^3.$$

Since $C$ is self-dual, the weight of every codeword is divisible by 3. We now ask for a linear code $C$ with no codewords of weight 3. For such a code the coefficient of $\theta_0^9\theta_1^3$ must be zero. This implies that

$$3\cdot 8-a=0,$$

hence $a=24$. Thus the weight enumerator of such a code is the polynomial

$$W_C(X_0,X_1)=X_0^{12}+264X_0^6X_1^6+440X_0^3X_1^9+24X_1^{12}.$$

Its expression in $E_4$ and $E_6^2$ is

$$W_C=\frac{5}{8}E_4^3+\frac{3}{8}E_6^2.$$

In fact, such a code exists: It is the extended ternary Golay code introduced in Sect. 5.2.

## 5.5 The Equation of the Tetrahedron and the Cube

We have seen in the proof of Proposition 5.4 that

$$\begin{aligned}\theta_0^4+8\theta_0\theta_1^3&=E_4,\\ \left(\theta_1^4-\theta_0^3\theta_1\right)^3&=-\frac{1}{4^3}\left(E_4^3-E_6^2\right),\\ \theta_0^6-20\theta_0^3\theta_1^3-8\theta_1^6&=E_6.\end{aligned}$$

The zeros of the homogeneous polynomial $e_T:=\theta_0^4+8\theta_0\theta_1^3$ in $\mathbb{P}_1(\mathbb{C})$ are the vertices of a tetrahedron as explained above. The zeros of the homogeneous polynomial $f_T:=4\left(\theta_1^4-\theta_0^3\theta_1\right)$ are the midpoints of the faces of this tetrahedron, the zeros of the homogeneous polynomial $k_T:=\theta_0^6-20\theta_0^3\theta_1^3-8\theta_1^6$ are the midpoints of the edges of this tetrahedron (always projected from the origin in $\mathbb{R}^3$ to the sphere $S^2$). We have the relation

$$e_T^3+f_T^3=k_T^2$$

This identity is called the *tetrahedral equation* (cf. also [92]).

Now there are two tetrahedra in a cube, as shown in Fig. 5.3. The homogeneous polynomials $e_T=\theta_0^4+8\theta_0\theta_1^3$ and $f_T=4\left(\theta_1^4-\theta_0^3\theta_1\right)$ are transformed into each other by the matrix

$$\begin{pmatrix}0&\sqrt{2}\\-\frac{1}{\sqrt{2}}&0\end{pmatrix}.$$

This transformation extends the symmetry group of the tetrahedron $G/\Gamma(3)$ to the symmetry group of the cube. The zeros of $e_T$ are vertices of one of the two tetrahedra in the cube, the zeros of $f_T$, which correspond to the midpoints of the faces of this tetrahedron, are the vertices of the other tetrahedron. So the 8 zeros of

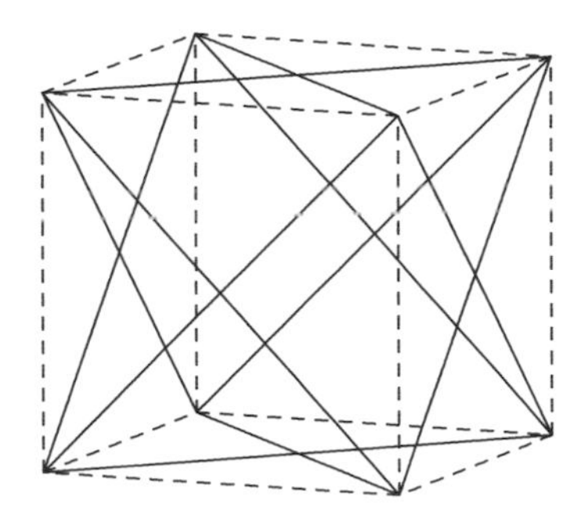

**Fig. 5.3** The two tetrahedra in a cube

$$e_C := e_T \cdot f_T$$

are the vertices of the cube. The zeros of

$$f_C := k_T = E_6$$

are the midpoints of the 6 faces of the cube. The polynomial

$$k_C := e_T^3 - f_T^3$$

is also invariant under the symmetry group of the cube. Its zeros are the midpoints of the 12 edges of the cube. By the tetrahedral equation we have

$$\begin{aligned} f_C^4 &= \left(e_T^3 + f_T^3\right)^2 = \left(e_T^3 - f_T^3\right)^2 + 4e_T^3 f_T^3 \\ &= k_C^2 + 4e_C^3 \end{aligned}$$

This yields the equation

$$f_C^4 = k_C^2 + 4e_C^3,$$

which is called the *equation of the cube*.

This equation can also be obtained in another way. Once again we consider the binary case, i.e., the case $p = 2$. In Sect. 2.9 we considered two functions $A$ and $B$ defined by

$$\begin{aligned} A &= \sum_{x \in 2\mathbb{Z}} q^{x^2/4}, \qquad q = e^{2\pi i z}, \\ B &= \sum_{x \in 2\mathbb{Z}+1} q^{x^2/4}. \end{aligned}$$

They played the roles of $\theta_0$ and $\theta_1$ in the binary case (cf. Proposition 2.11), and we saw that

$$E_4 = A^8 + 14A^4B^4 + B^8.$$

They are not usual modular forms, but they are modular forms of *half-integral weight* for $\Gamma(4)$. In fact $A$ and $B$ are modular forms of weight $\frac{1}{2}$ for the subgroup $\Gamma(4) \subset \mathrm{SL}_2(\mathbb{Z})$. We have $\Gamma(4) \subset G$, and $G/\Gamma(4)$ is isomorphic to the symmetry group of a cube. Analogously to the case $\Gamma(3)$, the group $\mathrm{SL}_2(\mathbb{Z})/\Gamma(4)$ is the *binary cube group*. The functions $A, B$ define a mapping

$$\frac{A}{B} : \overline{\mathbb{H}/\Gamma(4)} \longrightarrow \mathbb{P}_1(\mathbb{C}).$$

This mapping is again a bijection. The action of the group $G/\Gamma(4)$ on $\overline{\mathbb{H}/\Gamma(4)}$ corresponds to the action of the cube group on a cube lying inside $S^2 = \mathbb{P}_1(\mathbb{C})$.

Let $S^2$ be the unit sphere in $\mathbb{R}^3$ around the origin and consider a cube inscribed in $S^2$ with faces perpendicular to the coordinate axes. Consider the projection of this cube to the sphere $S^2$ from the origin. Again we identify this sphere with $\mathbb{P}_1(\mathbb{C})$ by the stereographical projection from the north pole to the equatorial plane. The homogeneous coordinates of $\mathbb{P}_1(\mathbb{C})$ are now denoted by $(A : B)$ with $\infty = (1 : 0)$. The homogeneous polynomial $e := E_4 = A^8 + 14A^4B^4 + B^8$ in the variables $A$ and $B$ has 8 zeros in $\mathbb{P}_1(\mathbb{C})$ which are the vertices of the cube, cf. Fig. 5.4.

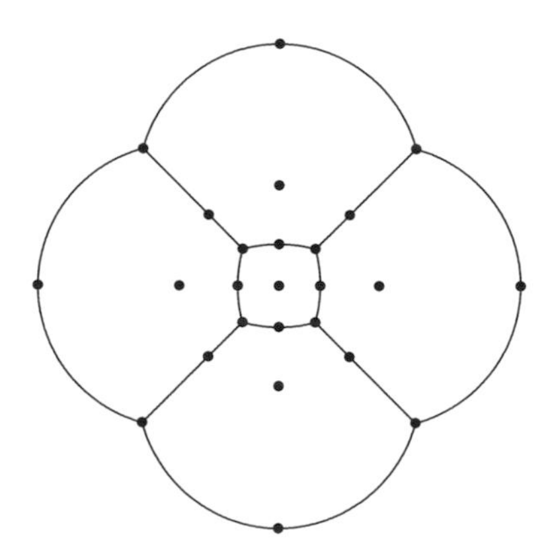

**Fig. 5.4** Stereographical projection of a cube

The midpoints of the faces of this cube are given as the zeros of

$$f := AB\left(A^4 - B^4\right).$$

We have seen in Sect. 2.9 that

$$f^4 = A^4B^4\left(A^4 - B^4\right)^4 = \frac{16}{1728}\left(E_4^3 - E_6^2\right).$$

Hence we get

$$108f^4 = E_4^3 - E_6^2 = e^3 - k^2,$$

with $k := E_6$, as before. This is, up to coefficients, again the equation of the cube. We also get

$$E_6 = A^{12} - 33A^8B^4 - 33A^4B^8 + B^{12}.$$

## 5.6 The Case $p = 5$: the Icosahedral Group

As a final application of Theorem 5.3, we consider the case $p = 5$. In this case, the field $k$ is the field $\mathbb{Q}(\sqrt{5})$. The ring of integers $\mathfrak{O}_k$ of $k$ consists of all linear combinations $a_0 + a_1(1+\sqrt{5})/2$ with $a_0, a_1 \in \mathbb{Z}$. The ideal $\mathfrak{p}$ is the ideal generated by $\sqrt{5}$ in $\mathfrak{O}_k$. The group $\Gamma(\mathfrak{p})$ is the group

$$\Gamma(\mathfrak{p}) = \left\{ \begin{pmatrix} \alpha & \beta \\ \gamma & \delta \end{pmatrix} \in \mathrm{SL}_2(\mathfrak{O}_k) \,\middle|\, \begin{matrix} \alpha \equiv \delta \equiv 1 \pmod{\sqrt{5}}, \\ \beta \equiv \gamma \equiv 0 \pmod{\sqrt{5}} \end{matrix} \right\}.$$

Because $-1 \notin \Gamma(\mathfrak{p})$, the subgroup $\Gamma(\mathfrak{p})$ can be regarded as a subgroup of the *Hilbert modular group* $G = \mathrm{SL}_2(\mathfrak{O}_k)/\{\pm 1\} = \mathrm{PSL}_2(\mathfrak{O}_k)$. The group $\Gamma(\mathfrak{p})$ acts freely on $\mathbb{H}^2$. The quotient $\mathbb{H}^2/\Gamma(\mathfrak{p})$ is called a *Hilbert modular surface* (see e.g. [33] and [26]). The functions $\theta_0, \theta_1$, and $\theta_2$ defined in Sect. 5.3 are Hilbert modular forms of weight 1 for $\Gamma(\mathfrak{p})$. They are symmetric, i.e., for all $z_1, z_2 \in \mathbb{H}$ we have

$$\theta_j(z_1, z_2) = \theta_j(z_2, z_1) \qquad \text{for } j = 0, 1, 2.$$

The factor group $G/\Gamma(\mathfrak{p})$ is isomorphic to $\mathrm{PSL}_2(\mathbb{F}_5)$ because $\mathfrak{O}_k/\mathfrak{p} \cong \mathbb{F}_5$. The group $\mathrm{PSL}_2(\mathbb{F}_5)$ is isomorphic to the alternating group $\mathscr{A}_5$. This is the symmetry group $I$ of the icosahedron and acts on the six axes of the icosahedron through its vertices in the same way as $\mathrm{PSL}_2(\mathbb{F}_5)$ acts on the six points of the projective line $\mathbb{P}_1(\mathbb{F}_5)$.

The icosahedral group $I$ is a subgroup of $\mathrm{SO}_3(\mathbb{R})$. It operates linearly on $\mathbb{R}^3$, where we take as standard coordinates the coordinates $x_0, x_1, x_2$, and thus also on $\mathbb{P}_2(\mathbb{R})$ and $\mathbb{P}_2(\mathbb{C})$. We shall consider the action on $\mathbb{P}_2(\mathbb{C})$. The invariant theory of this action was studied by Felix Klein [40]. We describe some of his results, following [34].

The first invariant polynomial considered by Klein is the polynomial

$$A := x_0^2 + x_1^2 + x_2^2,$$

which is invariant because it gives the equation of the sphere. This is an invariant polynomial of degree 2. Klein uses coordinates

$$A_0 = x_0, \qquad A_1 = x_1 + ix_2, \qquad A_2 = x_1 - ix_2.$$

In these coordinates, $A = A_0^2 + A_1A_2$, and this defines an invariant conic. The action of the icosahedral group $I$ on $\mathbb{P}_2(\mathbb{C})$ has exactly 6 fixed points with an isotropy group of order 10. These points are called *poles*. They correspond to the six pairs of opposite vertices of the icosahedron. Klein puts the icosahedron in such a position that the six poles are given by

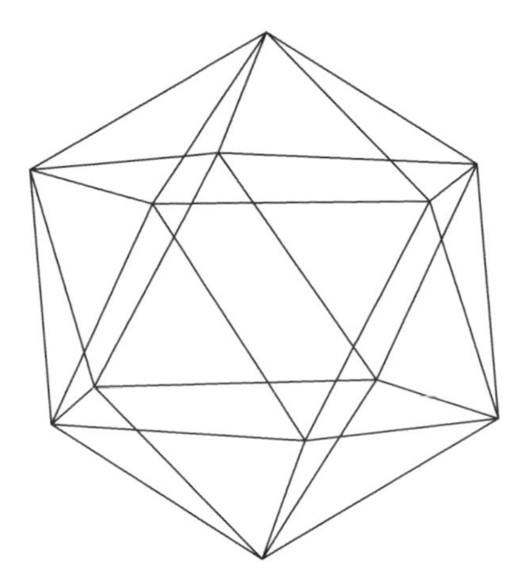

**Fig. 5.5** The icosahedron

$$(A_0,A_1,A_2) = \left(\frac{\sqrt{5}}{2},0,0\right),$$

$$(A_0,A_1,A_2) = \left(\frac{1}{2},\zeta^{\nu},\zeta^{-\nu}\right), \qquad \zeta = e^{2\pi i/5}, \quad 0 \leq \nu \leq 4.$$

The invariant curve $A = 0$ does not pass through the poles. Let $B$ be a homogeneous polynomial of degree 6 which vanishes in the poles. Let $C$ be a homogeneous polynomial of degree 10 which vanishes in the poles of higher order. Let $D$ be a homogeneous polynomial of degree 15 such that $D = 0$ is the union of the 15 lines connecting the six poles. Such polynomials exist and are unique up to a constant factor. The zero sets of these polynomials are complex-algebraic curves in $\mathbb{P}_2(\mathbb{C})$. The curves $B = 0$, $C = 0$, and $D = 0$ pass through the poles, and the poles are singular points of these curves. The singular point of $B = 0$ in a pole is a so called ordinary double point. This means that the picture of a corresponding *real* curve in a neighbourhood of a pole looks as in Fig. 5.6. The curve $C = 0$

**Fig. 5.6** Singular point of $B = 0$ in a pole (ordinary double point)

has a double cusp with separate tangents in each pole, i.e., the picture of a corresponding *real* curve in a neighbourhood of a pole looks as in Fig. 5.7. Klein gives explicit formulas for the polynomials $A$, $B$, $C$, and $D$. They generate the ring of all polynomials invariant

**Fig. 5.7** Singular point of $C = 0$ in a pole (double cusp with separate tangents)

under $I$. We list Klein's formulas:

$$\begin{aligned}
A &= A_0^2 + A_1A_2,\\
B &= 8A_0^4A_1A_2 - 2A_0^2A_1^2A_2^2 + A_1^3A_2^3 - A_0\left(A_1^5 + A_2^5\right),\\
C &= 320A_0^6A_1^2A_2^2 - 160A_0^4A_1^3A_2^3 + 20A_0^2A_1^4A_2^4 + 6A_1^5A_2^5\\
&\quad - 4A_0\left(A_1^5 + A_2^5\right)\left(32A_0^4 - 20A_0^2A_1A_2 + 5A_1^2A_2^2\right) + A_1^{10} + A_2^{10},\\
12D &= \left(A_1^5 - A_2^5\right)\left(-1024A_0^{10} + 3840A_0^8A_1A_2 - 3840A_0^6A_1^2A_2^2\right.\\
&\quad \left. + 1200A_0^4A_1^3A_2^3 - 100A_0^2A_1^4A_2^4 + A_1^5A_2^5\right)\\
&\quad + A_0\left(A_1^{10} - A_2^{10}\right)\left(352A_0^4 - 160A_0^2A_1A_2 + 10A_1^2A_2^2\right)\\
&\quad + \left(A_1^{15} - A_2^{15}\right).
\end{aligned}$$

These polynomials satisfy a relation $R(A,B,C,D) = 0$. Here $R(A,B,C,D)$ is given by

$$\begin{aligned}
R(A,B,C,D) &= -144D^2 - 1728B^5 + 720ACB^3\\
&\quad - 80A^2C^2B + 64A^3\left(5B^2 - AC\right)^2 + C^3.
\end{aligned}$$

According to Klein we have the following theorem.

**Theorem 5.6 (Klein).**

$$\mathbb{C}[A_0,A_1,A_2]^I = \mathbb{C}[A,B,C,D]/\left(R(A,B,C,D) = 0\right).$$

The conic $A = 0$ is isomorphic to $\mathbb{P}_1(\mathbb{C})$. When we restrict the action of $I$ to the conic $A = 0$, we get the action of the icosahedral group on $S^2 = \mathbb{P}_1(\mathbb{C})$. The curves $B = 0$, $C = 0$, and $D = 0$ intersect $A = 0$ transversally in 12, 20, and 30 points respectively. These points correspond to the 12 vertices, 20 midpoints of the faces, and 30 midpoints of the edges of the icosahedron (always projected from the origin of $\mathbb{R}^3$ to $S^2$). Putting $A = 0$ in the relation $R(A,B,C,D) = 0$ gives the *icosahedral equation* ([42], [41], [92], [35])

$$144k_I^2 = -1728e_I^5 + f_I^3,$$

with $k_I = D$, $e_I = B$, and $f_I = C$.

**Remark 5.5** If we consider $e_I$, $k_I$, and $f_I$ as complex variables, then this equation defines a hypersurface in $\mathbb{C}^3$ with an isolated singularity at the origin. In the same way the tetrahedral equation and the equation of the cube define isolated singularities of hypersurfaces in $\mathbb{C}^3$. For more information about these singularities see e.g. [52].

Now Hirzebruch has proved the following theorem [34]:

**Theorem 5.7 (Hirzebruch).** *The ring of symmetric Hilbert modular forms for $\Gamma(\mathfrak{p})$ equals $\mathbb{C}[A_0, A_1, A_2]$ where $A_0, A_1, A_2$ have weight* 1.

From Theorem 5.6 and Theorem 5.7 one gets the following corollary:

**Corollary 5.4.** *The ring of symmetric Hilbert modular forms for* $\mathrm{SL}_2(\mathfrak{O}_k)$ *of* even *weight equals* $\mathbb{C}[A,B,C]$.

Let $C \subset \mathbb{F}_5^2$ be the code of Example 5.3. This is a self-dual code. The Lee weight enumerator of $C$ is

$$W_C(X_0, X_1, X_2) = X_0^2 + 4X_1X_2.$$

By Theorem 5.3,

$$\theta_C = W_C(\theta_0, \theta_1, \theta_2) = \theta_0^2 + 4\theta_1\theta_2.$$

By Theorem 5.2, $\theta_C$ is a Hilbert modular form for $\mathrm{SL}_2(\mathfrak{O}_k)$ of weight 2. By Remark 5.4, it is symmetric. Therefore Corollary 5.4 implies that $\theta_C$ is a complex multiple of $A$. This shows that $\theta_0, \theta_1, \theta_2$ form a basis of the vector space of symmetric Hilbert modular forms for $\Gamma(\mathfrak{p})$ of weight 1.

Let

$$R = \begin{pmatrix} \alpha & \beta \\ \gamma & \delta \end{pmatrix} \in \mathrm{SL}_2(\mathfrak{O}),$$

and let $\theta$ be a symmetric Hilbert modular form of weight 1 for $\Gamma(\mathfrak{p})$. Then we set

$$(\theta|_1 R)(z_1, z_2) := (\sigma_1(\gamma)z_1 + \sigma_1(\delta))^{-1}(\sigma_2(\gamma)z_2 + \sigma_2(\delta))^{-1}\theta(\sigma_1(R)z_1, \sigma_2(R)z_2).$$

This is again a symmetric Hilbert modular form of weight 1 for $\Gamma(\mathfrak{p})$. This defines an action of $\mathrm{SL}_2(\mathfrak{O})$ on the space of Hilbert modular forms for $\Gamma(\mathfrak{p})$ of weight 1, hence on $\mathbb{C}\cdot\theta_0 \oplus \mathbb{C}\cdot\theta_1 \oplus \mathbb{C}\cdot\theta_2$. The group $\Gamma(\mathfrak{p})$ acts trivially. The action of $R$ is described by a $3\times 3$ matrix $\widetilde{R}$. By Formula (5.3) and Proposition 5.8 of Sect. 5.7 we get the following matrices for the specific elements of $\mathrm{SL}_2(\mathfrak{O})$:

$$\widetilde{S} = |_1 \begin{pmatrix} 0 & -1 \\ 1 & 0 \end{pmatrix} = -\frac{1}{\sqrt{5}} \begin{pmatrix} 1 & 2 & 2 \\ 1 & \zeta+\zeta^4 & \zeta^2+\zeta^3 \\ 1 & \zeta^2+\zeta^3 & \zeta+\zeta^4 \end{pmatrix},$$

$$\widetilde{T} = |_1 \begin{pmatrix} 1 & 1 \\ 0 & 1 \end{pmatrix} = \begin{pmatrix} 1 & 0 & 0 \\ 0 & \zeta^2 & 0 \\ 0 & 0 & \zeta^3 \end{pmatrix},$$

$$\widetilde{U} = \|_1 \begin{pmatrix} \zeta+\zeta^{-1} & 0 \\ 0 & -\zeta^2-\zeta^{-2} \end{pmatrix} = -\begin{pmatrix} 1 & 0 & 0 \\ 0 & 0 & 1 \\ 0 & 1 & 0 \end{pmatrix}.$$

We denote by $\overline{S}, \overline{T}, \overline{U}$ the images of $\widetilde{S}, \widetilde{T}, \widetilde{U}$ respectively in $G/\Gamma(\mathfrak{p}) \cong \mathrm{PSL}_2(\mathbb{F}_5) = I$. Then $\overline{S}$, $\overline{T}$ and $\overline{U}$ generate the icosahedral group $I$, and this is the 3-dimensional representation of $I$ considered by Klein [40, p. 349] (cf. also [42, p. 213]). Therefore we see that

$$A_0 = \theta_0, \quad A_1 = 2\theta_1, \quad A_2 = 2\theta_2.$$

**Exercise 5.2** Let $C \subset \mathbb{F}_5^n$ be a linear code with $C \subset C^\perp$. Derive from Theorem 5.3 and Proposition 5.7 the following identity for the Lee weight enumerator of $C$ and $C^\perp$:

$$W_{C^\perp}(X_0,X_1,X_2) = \frac{1}{5^m} W_C \begin{pmatrix} X_0 + 2X_1 + 2X_2, \\ X_0 + (\zeta+\zeta^4)X_1 + (\zeta^2+\zeta^3)X_2, \\ X_0 + (\zeta^2+\zeta^3)X_1 + (\zeta+\zeta^4)X_2 \end{pmatrix}$$

This is the *MacWilliams identity for Lee weight enumerators* of linear codes over $\mathbb{F}_5$ (cf. Theorem 2.7, Theorem 5.5, and [66, Chap. 5, §6, Theorem 12]).

**Exercise 5.3** Let $C \subset \mathbb{F}_2^n$ be a self-dual code. Using Exercise 5.2, show that the Lee weight enumerator of $C$ is invariant under the transformations $\widetilde{S}$, $\widetilde{T}$, and $\widetilde{U}$ of the $(X_0,X_1,X_2)$-space.

Putting everything together we get:

**Corollary 5.5 (Gleason and Pierce; Sloane).** *The Lee weight enumerator of a self-dual code over $\mathbb{F}_5$ is a polynomial in the Klein invariants $A, B, C$.*

This corollary was first proved independently by A. M. Gleason and J. N. Pierce (unpublished) and N. J. A. Sloane (cf. [64]).

**Example 5.5** Let $C \subset \mathbb{F}_5^6$ be the self-dual code with generator matrix

$$\begin{pmatrix} 0 & 0 & 1 & 1 & 2 & -2 \\ 2 & -2 & 0 & 0 & 1 & 1 \\ 1 & 1 & 2 & -2 & 0 & 0 \end{pmatrix}$$

considered in Sect. 5.3. By Corollary 5.5 the Lee weight enumerator of $C$ has the form

$$A^3 + aB$$

for some $a \in \mathbb{C}$. One can easily see that the coefficient of $X_0X_1^5$ is equal to 12. This implies that

$$a = -\frac{3}{8}.$$

Hence we obtain as the Lee weight enumerator of $C$:

$$W_C(X_0,X_1,X_2) = A^3 - \frac{3}{8}B = X_0^6 + 12X_0(X_1^5 + X_2^5) + 60X_0^2X_1^2X_2^2 + 40X_1^3X_2^3.$$

## 5.7 Theta Functions as Hilbert Modular Forms (by N.-P. Skoruppa)

In this section we give the proofs of Theorem 5.1 and Theorem 5.2. We indicate how one can verify that certain theta functions are Hilbert modular forms. The method is entirely analogous to the method presented in Sect. 3.1 for theta functions of integral lattices. It goes back to Hecke, Schoeneberg, Kloosterman (and other people). Here we consider the case of a lattice over the integers of a totally real algebraic number field. It seems that this case was first treated by M. Eichler [24]. We follow the note [83].

We start with some general remarks concerning lattices over integers of a totally real algebraic number field. Let $K$ be an algebraic number field of finite degree $r = [K : \mathbb{Q}]$. Let $\sigma_1, \dots, \sigma_r : K \to \mathbb{C}$ be the different embeddings of $K$ into $\mathbb{C}$ with $\sigma_1 = \mathrm{id}$. Let $y \in K$. Recall that

$$N(y) := \prod_{i=1}^{r} \sigma_i(y)$$

is the norm of $y$, and

$$\mathrm{Tr}(y) := \sum_{i=1}^{r} \sigma_i(y)$$

is the trace of $y$. These are rational numbers. Let $\mathfrak{O}$ be the ring of integers of $K$. A *fractional ideal* in $K$ is a finitely generated $\mathfrak{O}$-module in $K$. Let $\mathfrak{a}$ and $\mathfrak{b}$ be fractional ideals in $K$. Then one defines their *product* $\mathfrak{a} \cdot \mathfrak{b}$ as the $\mathfrak{O}$-submodule of $K$ generated by all products $y \cdot y'$ where $y \in \mathfrak{a}$ and $y' \in \mathfrak{b}$. For a nonzero fractional ideal $\mathfrak{a}$ we define

$$\mathfrak{a}^{-1} := \{y \in K \mid y\mathfrak{a} \subset \mathfrak{O}\}.$$

Then one can easily show that $\mathfrak{a}^{-1}$ is again a fractional ideal, and that $\mathfrak{a} \cdot \mathfrak{a}^{-1} = \mathfrak{O}$. So the nonzero fractional ideals in $K$ form a *group* with respect to the above multiplication. The neutral element is the ideal $\mathfrak{O}$.

It will follow from a more general result proved below that a fractional ideal is a free $\mathbb{Z}$-module of rank $r$.

For a fractional ideal $\mathfrak{a}$ one defines the *dual ideal* $\mathfrak{a}^*$ as

$$\mathfrak{a}^* := \{y \in K \mid \mathrm{Tr}(y\mathfrak{a}) \subset \mathbb{Z}\}.$$

This is also a fractional ideal. In particular one can consider the fractional ideal $\mathfrak{D} := (\mathfrak{O}^*)^{-1}$. It is integral, i.e., $\mathfrak{D} \subset \mathfrak{O}$, as follows immediately from $\mathfrak{O}^* \supset \mathfrak{O}$. The ideal $\mathfrak{D}$ is called the *different* of $K$ over $\mathbb{Q}$.

**Proposition 5.5.** *Let* $\mathfrak{a}$ *be a fractional ideal. Then*

$$\mathfrak{a}^* = \mathfrak{D}^{-1}\mathfrak{a}^{-1}.$$

*Proof.* We have

$$\mathrm{Tr}\left(\mathfrak{D}^{-1}\mathfrak{a}^{-1}\mathfrak{a}\right) = \mathrm{Tr}\left(\mathfrak{O}^*\mathfrak{O}\right) \subset \mathbb{Z},$$

whence $\mathfrak{D}^{-1}\mathfrak{a}^{-1} \subset \mathfrak{a}^*$. For the converse, note that

$$\mathrm{Tr}(\mathfrak{a}^*\mathfrak{a}\mathfrak{O}) \subset \mathbb{Z},$$

whence $\mathfrak{a}^*\mathfrak{a} \subset \mathfrak{O}^* = \mathfrak{D}^{-1}$. This proves the proposition. □

We now make the general assumption that $K$ is *totally real*, i.e., $\sigma_j(K) \subset \mathbb{R}$ for all $1 \leq j \leq r$. We consider a vector space $V$ over $K$ of finite dimension $n = \dim_K V$ provided with a *totally positive definite scalar product* " $\cdot$ " (i.e., $\cdot : V \times V \to K$ is a bilinear symmetric mapping, and $\sigma_j(v \cdot v) > 0$ for all $1 \leq j \leq r$ and all $v \in V \setminus \{0\}$).

A *$K$-lattice $\Gamma$* in $V$ is a finitely generated $\mathfrak{O}$-submodule $\Gamma$ of $V$ which contains a $K$-basis of $V$.

**Example 5.6** Let $V = K$ and " $\cdot$ " be the multiplication in $K$. A nonzero fractional ideal in $K$ is a $K$-lattice in $V$.

**Example 5.7** Let $p$ be an odd prime number, $\zeta = e^{2\pi i/p}$. Let $K = \mathbb{Q}(\zeta + \zeta^{-1})$ and $V := \mathbb{Q}(\zeta)$. Then $V$ is a 2-dimensional $K$-vector space. For $v, w \in V$ let

$$v \cdot w := \frac{v\overline{w} + \overline{v}w}{p}.$$

This defines a totally positive definite scalar product " $\cdot$ " on $V$. Let $\Gamma$ be the principal ideal of $\mathfrak{O}_V$ generated by the element $1 - \zeta \in \mathfrak{O}_V$. Then $\Gamma$ is a $K$-lattice in $V$.

More generally let $V := \mathbb{Q}(\zeta)^n$. Then $V$ is a $2n$-dimensional $K$-vector space. Define the scalar product " $\cdot$ " on $V$ as the sum of the scalar products of the individual coordinates as defined before. This scalar product is totally positive definite. Let $C \subset \mathbb{F}_p^n$ be a self-dual code, and let $\Gamma_C \subset V$ be the lattice constructed from $C$ in Sect. 5.2. Then $\Gamma_C$ is a $K$-lattice.

For a $K$-lattice $\Gamma$ in $V$ we define the *dual lattice* $\Gamma^*$ by

$$\Gamma^* := \{v \in V \mid \mathrm{Tr}(v \cdot \Gamma) \subset \mathbb{Z}\}.$$

**Proposition 5.6.** (i) *A $K$-lattice in $V$ is a free $\mathbb{Z}$-module of rank $r \cdot n$.*

(ii) *The set $\Gamma^*$ is a $K$-lattice in $V$.*

*Proof.* (i) A $K$-lattice in $V$ is in any case a free $\mathbb{Z}$-module of finite rank, since it is finitely generated and torsion free. We show that any two $K$-lattices $\Gamma_1$ and $\Gamma_2$ in $V$ are *commensurable*, i.e., $\Gamma_1 \cap \Gamma_2$ has finite index in both $\Gamma_1$ and $\Gamma_2$. Let $\Gamma_1$ and $\Gamma_2$ be $K$-lattices in $V$. Then there exists a number $N \in \mathbb{Z}$, $N > 0$, such that $N\Gamma_1 \subset \Gamma_2$. For let $\Gamma_1 = \mathfrak{O}v_1 + \ldots + \mathfrak{O}v_s$, $\Gamma_2 = \mathfrak{O}w_1 + \ldots + \mathfrak{O}w_t$ for appropriate $v_j, w_k \in V$, where $\{v_1, \ldots, v_s\}$ and $\{w_1, \ldots, w_t\}$ are $K$-bases of $V$. Then $v_j = \sum_k y_{jk} w_k$ for appropriate $y_{jk} \in K$. There exists $N \in \mathbb{Z}$, $N > 0$ with $Ny_{jk} \in \mathfrak{O}$ for all $1 \leq j \leq s$, $1 \leq k \leq t$. Thus $Nv_j \in \Gamma_2$ for all $1 \leq j \leq s$, whence $N\Gamma_1 \subset \Gamma_2$. Clearly $[\Gamma_1 : N\Gamma_1] < \infty$. Together with $N\Gamma_1 \subset \Gamma_1 \cap \Gamma_2 \subset \Gamma_1$, this implies $[\Gamma_1 : \Gamma_1 \cap \Gamma_2] < \infty$. Analogously, one can derive $[\Gamma_2 : \Gamma_1 \cap \Gamma_2] < \infty$.

Since any two $K$-lattices in $V$ are commensurable, each $K$-lattice in $V$ is a free $\mathbb{Z}$-module of rank $r \cdot n$, if only one $K$-lattice in $V$ has this property. But evidently $\Gamma := \mathfrak{O}v_1 + \ldots + \mathfrak{O}v_n$, where $\{v_1, \ldots, v_n\}$ is a $K$-basis of $V$, has rank $r \cdot n$, since $\mathfrak{O}$ is a free $\mathbb{Z}$-module of rank $r$.

(ii) Obviously, $\Gamma^*$ is an $\mathfrak{O}$-module. Write $\Gamma = \mathbb{Z}e_1 + \ldots + \mathbb{Z}e_{rn}$, where $\{e_1, \ldots, e_{rn}\}$ is a basis of $V$ over $\mathbb{Q}$. Then $\Gamma^* = \mathbb{Z}e_1^* + \ldots + \mathbb{Z}e_{rn}^*$ where the $e_j^*$ are determined by the condition $\mathrm{Tr}\left(e_j^* \cdot e_k\right) = \delta_{jk}$ (Kronecker symbol). Therefore $\Gamma^*$ is finitely generated and contains a basis of $V$ over $K$. Hence $\Gamma^*$ is a $K$-lattice in $V$. This concludes the proof of Proposition 5.6. □

Let $\Gamma$ be a $K$-lattice in $V$. Let $\{e_1, \ldots, e_{rn}\}$ be a $\mathbb{Z}$-basis of $\Gamma$ as in the proof of the previous proposition. Then

$$\Delta(\Gamma) := \det(\mathrm{Tr}(e_i \cdot e_j))$$

is called the *discriminant* of $\Gamma$. It does not depend on the choice of the basis $\{e_1, \ldots, e_{rn}\}$.

We now fix a $K$-lattice $\Gamma$ in $V$. For a given $v_0 \in V$ we define the *theta function*

$$\theta_{v_0+\Gamma}(z_1, \ldots, z_r) := \sum_{v \in v_0+\Gamma} e^{\pi i \mathrm{Tr}(zv^2)},$$

where $z_1, \ldots z_r \in \mathbb{H}$, and

$$\mathrm{Tr}(zv^2) := \sum_{j=1}^{r} z_j \sigma_j(v \cdot v).$$

**Proposition 5.7.** *The series $\theta_{v_0+\Gamma}$ is absolutely uniformly convergent in each subset of $\mathbb{H}^r$ of the form* $\{(z_1, \ldots, z_r) \in \mathbb{H}^r \mid \mathrm{Im}\, z_j > y_0$ for $1 \leq j \leq n\}$ $(y_0 \in \mathbb{R},\ y_0 > 0)$. *In particular $\theta_{v_0+\Gamma}$ is holomorphic in $\mathbb{H}^r$. One has*

$$\theta_{v_0+\Gamma}\left(-\frac{1}{z_1}, \ldots, -\frac{1}{z_r}\right) = \left(\frac{z_1}{i}\right)^{n/2} \cdot \ldots \cdot \left(\frac{z_r}{i}\right)^{n/2} \Delta(\Gamma)^{-1/2} \sum_{v \in \Gamma^*} e^{2\pi i \mathrm{Tr}(v \cdot v_0)} e^{\pi i \mathrm{Tr}(zv^2)}.$$

*(Here and in the sequel we adopt the convention $w^s := e^{s(\log|w| + i \arg w)}$ with $-\pi < \arg w \leq +\pi$ for $w, s \in \mathbb{C}$, $w \neq 0$.)*

The proof of this proposition will be analogous to the proof of Proposition 3.1. We shall apply the Poisson summation formula. For that purpose we need the following lemma.

**Lemma 5.6.** *Let $Z$ be a symmetric complex $r \times r$-matrix, $\mathrm{Im}\, Z > 0$ (i.e., $Z = X + iY$ with real $X, Y$ and $Y$ positive definite). Let $x_0 \in \mathbb{R}^r$. Then*

$$\int_{\mathbb{R}^r} e^{-\pi i (x+x_0) \cdot Z^{-1}(x+x_0)} e^{-2\pi i x \cdot y}\, dx = \pm \det\left(\frac{Z}{i}\right)^{1/2} e^{2\pi i x_0 \cdot y} e^{\pi i y \cdot Z y}.$$

*Proof.* If one substitutes $x - x_0 \mapsto x$ in the integral on the left-hand side, then one observes that it suffices to proof the formula for $x_0 = 0$. We can write $Z = X + iY$ with real symmetric matrices $X$, $Y$ and $Y$ positive definite. It is well-known that $X$ and $Y$ can be simultaneously diagonalized. Thus there is a $T \in \mathrm{GL}_r(\mathbb{R})$ such that

$$T^t Z T = \begin{pmatrix} z_1 & & 0 \\ & \ddots & \\ 0 & & z_r \end{pmatrix}$$

for appropriate $z_j \in \mathbb{C}$, $\operatorname{Im} z_j > 0$. If we set $y = Tw$, $T^t x = v$ in the formula of Lemma 5.6, then this formula becomes

$$\int_{\mathbb{R}^r} e^{-\pi\iota\left(\frac{v_1^2}{z_1}+\ldots+\frac{v_r^2}{z_r}\right)} e^{-2\pi i(v_1w_1+\ldots+v_rw_r)} \frac{1}{|\det T|} dv_1 \ldots dv_r$$

$$= \pm\det\left(\frac{Z}{i}\right)^{1/2} e^{\pi i\left(z_1w_1^2+\ldots+z_rw_r^2\right)},$$

where $v = (v_1, \ldots, v_r)$, $w = (w_1, \ldots, w_r)$. From this identity we can deduce on the one hand that the integral in the lemma is defined, and on the other hand that it suffices to prove the formula for the case $r = 1$. But for $r = 1$ the formula reduces to the first entry of Table 3.1 in Sect. 3.1. This proves Lemma 5.6. □

*Proof of Proposition 5.7.* For each $1 \leq j \leq r$, $\mathbb{R}$ becomes a $K$-module via

$$K \times \mathbb{R} \ni (x, \rho) \longmapsto \sigma_j(x) \cdot \rho;$$

we denote this $K$-module by $K_j$. Let $V_j := V \otimes_K K_j$. Then $V_j$ is a real vector space of dimension $n$, and $V \subset V_j$ via $V \ni v \mapsto v \otimes 1$. We extend the scalar product

$$V \times V \longrightarrow K_j, \qquad (v, w) \longmapsto \sigma_j(v \cdot w)$$

to a scalar product on $V_j$ in an $\mathbb{R}$-bilinear way. This turns $V_j$ into a Euclidean vector space. Let $L_j$ be an isomorphism of this Euclidean vector space $V_j$ with the standard Euclidean vector space $\mathbb{R}^n$, i.e., $L_j : V_j \to \mathbb{R}^n$ is an isomorphism with $\sigma_j(v \cdot w) = L_j(v) \cdot L_j(w)$ for all $v, w \in V$.

We now consider the embedding

$$\begin{aligned} L\colon V &\longrightarrow \mathbb{R}^{rn} \\ v &\longmapsto (L_1(v), \ldots, L_r(v)). \end{aligned}$$

Then one can easily see that $L(\Gamma)$ is a lattice in $\mathbb{R}^{rn}$, that $L(\Gamma^*) = L(\Gamma)^*$, and that

$$\theta_{v_0+\Gamma}(z_1, \ldots, z_r) = \sum_{\xi \in L(v_0)+L(\Gamma)} e^{\pi i \xi \cdot Z(\xi)}$$

with

$$Z = \begin{pmatrix} z_1E_n & & 0 \\ & \ddots & \\ 0 & & z_rE_n \end{pmatrix},$$

where $E_n$ is the $n \times n$ unit matrix.

For $\xi \in \mathbb{R}^{rn}$ and $(z_1, \ldots, z_r) \in \mathbb{H}^r$ with $\operatorname{Im} z_j > y_0$ for $1 \leq j \leq r$ one now has

$$\left|e^{\pi i \xi \cdot Z(\xi)}\right| < e^{-\pi y_0 \xi \cdot \xi}.$$

This implies the first assertion of Proposition 5.7. The claimed transformation formula is now a simple consequence of the formula of Lemma 5.6. This proves Proposition 5.7. □

From now on we make the general additional assumption that $\Gamma$ is *integral* (i.e., $\mathrm{Tr}(v\cdot w)\in\mathbb{Z}$ for all $v,w\in\Gamma$) and *even* (i.e., $\mathrm{Tr}(\mathfrak{O}v^2)\in 2\mathbb{Z}$ for all $v\in\Gamma$). Then $\Gamma\subset\Gamma^*$ and $\Delta(\Gamma)=[\Gamma^*:\Gamma]$. Also, for $v,w\in\Gamma^*$ the expressions $e^{\pi i\mathrm{Tr}(v^2)}$ and $e^{2\pi i\mathrm{Tr}(v\cdot w)}$ only depend on $v$ and $w$ modulo $\Gamma$. We also assume from now on that $n=\dim_K V$ is even. Let $k=\frac{n}{2}$.

We now consider the operation of $\mathrm{SL}_2(\mathfrak{O})$ on $\mathbb{H}^r$ defined in Sect. 5.3. For a function $f:\mathbb{H}^r\longrightarrow\mathbb{C}$, $k=\frac{n}{2}$, and

$$A=\begin{pmatrix}\alpha&\beta\\ \gamma&\delta\end{pmatrix}\in\mathrm{SL}_2(\mathfrak{O})$$

we set

$$(f\mid_k A)(z_1,\dots,z_r):=\prod_{j=1}^{r}(\sigma_j(\gamma)z_j+\sigma_j(\delta))^{-k}f(\sigma_1(A)z_1,\dots,\sigma_r(A)z_r).$$

This defines an operation of $\mathrm{SL}_2(\mathfrak{O})$ on the set of all functions $f:\mathbb{H}^r\to\mathbb{C}$. Using this operation we may rewrite the formula of Proposition 5.7 as

$$\theta_{v_0+\Gamma}\mid_k\begin{pmatrix}0&-1\\ 1&0\end{pmatrix}=i^{-kr}\Delta(\Gamma)^{-1/2}\sum_{w_0\in\Gamma^*/\Gamma}e^{2\pi i\mathrm{Tr}(v_0\cdot w_0)}\theta_{w_0+\Gamma}. \tag{5.3}$$

Here and in the sequel "$w_0\in\Gamma^*/\Gamma$" means that $w_0$ runs over a complete system of representatives for $\Gamma^*/\Gamma$.

**Proposition 5.8.** *The group* $\mathrm{SL}_2(\mathfrak{O})$ *leaves the span of all* $\theta_{v_0+\Gamma}$ $(v_0\in\Gamma^*)$ *invariant. More precisely we have the following formulas. Let* $v\in\Gamma^*$ *and* $A=\begin{pmatrix}\alpha&\beta\\ \gamma&\delta\end{pmatrix}\in\mathrm{SL}_2(\mathfrak{O})$. *Then we have (with* $k=\frac{n}{2}$*)*

$$\theta_{v+\Gamma}\mid_k A=i^{-kr}N(\gamma)^{-k}\Delta(\Gamma)^{-\frac{1}{2}}\cdot$$
$$\cdot\sum_{w\in\Gamma^*/\Gamma}\left(e^{-\pi i\mathrm{Tr}(2\beta vw+\beta\delta w^2)}\sum_{\substack{u\in\Gamma^*/\gamma\Gamma\\ u\equiv v+\delta w\,(\Gamma)}}e^{\pi i\mathrm{Tr}\left(\frac{\alpha}{\gamma}u^2\right)}\right)\theta_{w+\Gamma}$$

*if* $\gamma\neq 0$, *and*

$$\theta_{v+\Gamma}\mid_k A=N(\delta)^{-k}e^{\pi i\mathrm{Tr}(\alpha\beta v^2)}\theta_{\alpha v+\Gamma}$$

*if* $\gamma=0$.

The proof of Proposition 5.8 is completely analogous to the proof of Proposition 3.2: the case $\gamma=0$ is obvious and for $\gamma\neq 0$ one writes

$$\sigma_j(A)z_j=\frac{\sigma_j(\alpha)}{\sigma_j(\gamma)}-\frac{1}{\sigma_j(\gamma)(\sigma_j(\gamma)z_j+\sigma_j(\delta))}$$

and applies (5.3).

Now let

$$\mathfrak{L} := \left\{ x \in \mathfrak{O} \,\middle|\, \mathrm{Tr}\left( x\mathfrak{O}\frac{v^2}{2} \right) \subset \mathbb{Z} \quad \text{for all } v \in \Gamma^* \right\}.$$

We call $\mathfrak{L}$ the *level of* $\Gamma$. In the case when $K = \mathbb{Q}$,

$$\mathfrak{L} = \left\{ n \in \mathbb{Z} \,\middle|\, n\frac{v^2}{2} \in \mathbb{Z} \quad \text{for all } v \in \Gamma^* \right\} = (N) \subset \mathbb{Z},$$

for some $N \in \mathbb{Z}$, $N > 0$. In the case of an integral lattice $\Gamma$ we defined this number $N$ to be the level of $\Gamma$. It will follow from the next proposition that $\mathfrak{L}$ is an ideal in $\mathfrak{O}$.

**Proposition 5.9.** (i) *Let* $\mathfrak{N} := \gcd(\frac{v^2}{2} \mid v \in \Gamma^*)$, *i.e., let* $\mathfrak{N}$ *be the* $\mathfrak{O}$*-submodule of* $K$ *generated by the elements* $\frac{v^2}{2}$ *for* $v \in \Gamma^*$. *Then* $\mathfrak{N}$ *is a fractional ideal in* $K$.

(ii) *For the level* $\mathfrak{L}$ *of* $\Gamma$ *one has* $\mathfrak{L} = \mathfrak{N}^* = \mathfrak{N}^{-1}\mathfrak{D}^{-1}$ *and* $\mathfrak{L}\Gamma^* \subset \Gamma$.

(iii) *Let* $n(\mathfrak{L}) := [\mathfrak{O} : \mathfrak{L}]$ *(norm of* $\mathfrak{L}$*) and let* $e$ *be the exponent of the finite abelian group* $\Gamma^*/\Gamma$ *(i.e., the smallest number* $e$ *such that* $ev = 0$ *for all* $v \in \Gamma^*/\Gamma$*). Then* $e \mid n(\mathfrak{L})$. *Moreover,* $2e \in \mathfrak{L}$, *and* $e \in \mathfrak{L}$ *if* $e$ *is odd.*

(iv) *One has* $\mathfrak{L} = \mathfrak{O}$ *if and only if* $\Gamma = \Gamma^*$.

*Proof.* (i) Let $\Gamma^* = \mathfrak{O}v_1 + \ldots + \mathfrak{O}v_t$ for appropriate $v_1, \ldots, v_t \in \Gamma^*$. Then

$$\mathfrak{N} = \sum_j \mathfrak{O}\frac{v_j^2}{2} + \sum_{j,k} \mathfrak{O}v_j \cdot v_k.$$

Hence $\mathfrak{N}$ is a finitely generated $\mathfrak{O}$-module in $K$, i.e., a fractional ideal.

(ii) Evidently $\mathfrak{L} \subset \{x \in K \mid \mathrm{Tr}(x\mathfrak{N}) \subset \mathbb{Z}\} = \mathfrak{N}^*$. For the converse, it suffices to show that $\mathfrak{N}^* \subset \mathfrak{O}$. We first show that $\mathfrak{N}^*\Gamma^* \subset \Gamma$: Let $y \in \mathfrak{N}^*$, $v, w \in \Gamma^*$. Then

$$\mathrm{Tr}(y(v \cdot w)) = \mathrm{Tr}\left( y\left( \frac{(v+w)^2}{2} - \frac{v^2}{2} - \frac{w^2}{2} \right) \right) \in \mathbb{Z},$$

whence $y\Gamma^* \subset \Gamma^{**} = \Gamma$. Now $\mathfrak{N}^*\Gamma^* \subset \Gamma$ implies that $\mathfrak{N}^* \subset \mathfrak{O}$: Let $e_1, \ldots, e_{rn}$ be a $\mathbb{Z}$-basis of $\Gamma$, $y \in \mathfrak{N}^*$. Then $ye_j = \sum_k t_{jk}e_k$ with $t_{jk} \in \mathbb{Z}$, since $\mathfrak{N}^*\Gamma^* \subset \Gamma$. This implies $\det(t_{ij} - y\delta_{ij}) = 0$. But $\det(t_{ij} - y\delta_{ij})$ is a monic polynomial in $\mathbb{Z}[y]$ of degree $rn$. Thus $y \in \mathfrak{O}$.

(iii) One has $n(\mathfrak{L}) \in \mathfrak{L}$; by (ii) $n(\mathfrak{L}) \cdot \Gamma^* \subset \Gamma$, whence $e \mid n(\mathfrak{L})$. Let $v \in \Gamma^*$. Then $\mathrm{Tr}\left(2e\mathfrak{O}\frac{v^2}{2}\right) = \mathrm{Tr}(\mathfrak{O}ev \cdot v) \in \mathbb{Z}$, since $ev \in \Gamma$ and $v \in \Gamma^*$. Thus $2e \in \mathfrak{L}$. The statement $e \in \mathfrak{L}$ for odd $e$ is left to the reader.

(iv) follows from (iii). This finishes the proof of Proposition 5.9. □

Since $\mathfrak{L}$ is an ideal in $\mathfrak{O}$, we can define subgroups $\Gamma_0(\mathfrak{L})$, $\Gamma_1(\mathfrak{L})$, and $\Gamma(\mathfrak{L})$ of $\mathrm{SL}_2(\mathfrak{O})$ as follows:

$$\Gamma_0(\mathfrak{L}) := \left\{ \begin{pmatrix} \alpha & \beta \\ \gamma & \delta \end{pmatrix} \in \mathrm{SL}_2(\mathfrak{O}) \,\middle|\, \gamma \equiv 0 \pmod{\mathfrak{L}} \right\},$$

$$\Gamma_1(\mathfrak{L}) := \left\{ \begin{pmatrix} \alpha & \beta \\ \gamma & \delta \end{pmatrix} \in \mathrm{SL}_2(\mathfrak{O}) \,\middle|\, \begin{array}{l} \alpha \equiv \delta \equiv 1 \pmod{\mathfrak{L}}, \\ \gamma \equiv 0 \pmod{\mathfrak{L}} \end{array} \right\},$$
$$\Gamma(\mathfrak{L}) := \left\{ \begin{pmatrix} \alpha & \beta \\ \gamma & \delta \end{pmatrix} \in \mathrm{SL}_2(\mathfrak{O}) \,\middle|\, \begin{array}{l} \alpha \equiv \delta \equiv 1 \pmod{\mathfrak{L}}, \\ \gamma \equiv \beta \equiv 0 \pmod{\mathfrak{L}} \end{array} \right\}.$$

As in Sect. 3.1, Corollary 3.1, we have as an immediate consequence of Proposition 5.8:

**Corollary 5.6.** *Let* $A = \begin{pmatrix} \alpha & \beta \\ \gamma & \delta \end{pmatrix} \in \Gamma_0(\mathfrak{L})$, $v \in \Gamma^*$. *Then*

$$\theta_{v+\Gamma} \mid_k A = \varepsilon(A) e^{\pi i \mathrm{Tr}(\alpha\beta v^2)} \theta_{\alpha v + \Gamma},$$

*where*

$$\varepsilon(A) = \begin{cases} i^{-kr} N(\gamma)^{-k} \Delta(\Gamma)^{-1/2} \sum\limits_{u \in \Gamma/\gamma\Gamma} e^{\pi i \mathrm{Tr}\left(\frac{\alpha}{\gamma} u^2\right)} & \text{for } \gamma \neq 0, \\ N(\delta)^{-k} & \text{for } \gamma = 0. \end{cases}$$

Moreover, similar to the proof of Theorem 3.2 in Sect. 3.1 one can show the following proposition.

**Proposition 5.10.** *There exists a character* $\chi : (\mathfrak{O}/\mathfrak{L})^\times \to \{\pm 1\}$ *(*$(\mathfrak{O}/\mathfrak{L})^\times$ *multiplicative subgroup of* $\mathfrak{O}/\mathfrak{L}$*) such that*

$$\varepsilon \begin{pmatrix} \alpha & \beta \\ \gamma & \delta \end{pmatrix} = \chi(\delta \bmod \mathfrak{L}) \quad \text{for } \begin{pmatrix} \alpha & \beta \\ \gamma & \delta \end{pmatrix} \in \Gamma_0(\mathfrak{L}).$$

*For* $p \in \mathbb{Z}$, $\gcd(p, \mathfrak{L}) = 1$, *p prime, one has*

$$\chi(p \bmod \mathfrak{L}) = \left( \frac{(-1)^{rn/2} \Delta(\Gamma)}{p} \right) \qquad \text{(Legendre symbol)}.$$

From this and Corollary 5.6 we then obtain the analogue of Theorem 3.2:

**Theorem 5.8.** *We have for* $v \in \Gamma^*$

$$\begin{aligned} \theta_{v+\Gamma} \mid_k A &= \theta_{v+\Gamma} && \text{for } A \in \Gamma(\mathfrak{L}), \\ \theta_\Gamma \mid_k A &= \chi(\delta \bmod \mathfrak{L}) \theta_\Gamma && \text{for } A = \begin{pmatrix} \alpha & \beta \\ \gamma & \delta \end{pmatrix} \in \Gamma_0(\mathfrak{L}), \end{aligned}$$

*where* $\chi : (\mathfrak{O}/\mathfrak{L})^\times \to \{\pm 1\}$ *is the character of Proposition 5.10.*

Theorems 5.1 and 5.2 are obvious consequences of Theorem 5.8.

**Example 5.7 (continued)** We consider the lattice $\Gamma = (1-\zeta)$ in $V = \mathbb{Q}(\zeta)$ of Example 5.7. We have seen in Sect. 5.1 that $\Gamma^* = \mathfrak{O}_V$. Therefore

$$\Delta(\Gamma) = [\Gamma^* : \Gamma] = |\mathfrak{O}_V/(1-\zeta)| = p.$$

Using Proposition 5.9, one can show that

$$\mathfrak{L} = \left((1-\zeta)(1-\overline{\zeta})\right) \subset \mathfrak{O}_K,$$

so $\mathfrak{L} = \mathfrak{p}$ in the notation of Sect. 5.3. Therefore Theorem 5.1 follows from Theorem 5.8.

For the lattice $\Gamma_C$ of Example 5.7, $\Gamma_C^* = \Gamma_C$, so $\mathfrak{L} = \mathfrak{O}$ by Proposition 5.9. Thus Theorem 5.8 implies Theorem 5.2.

# References


1. L. V. Ahlfors. *Complex Analysis*. McGraw-Hill, New York etc., second edition, 1966.
2. A. E. Brouwer, A. M. Cohen, and A. Neumaier. *Distance-Regular Graphs*. Springer, Berlin etc., 1989.
3. A. Borel and J. de Siebenthal. Les sous-groupes fermés de rang maximum des groupes de Lie clos. *Comm. Math. Helv.*, 23:200–221, 1949.
4. M. Broué and M. Enguehard. Polynômes des poids de certains codes et fonctions thêta de certains réseaux. *Ann. scient. Ec. Norm. Sup.*, 5:157–181, 1972.
5. R. E. Borcherds. The Leech lattice. *Proc. R. Soc. Lond.*, A 398:365–376, 1985.
6. N. Bourbaki. *Groupes et Algèbres de Lie, Ch. 4,5 et 6*. Hermann, Paris, 1968.
7. A. E. Brouwer. The Witt designs, Golay codes and Mathieu groups. Unpublished manuscript, Eindhoven University of Technology, Eindhoven, 1982?
8. P. J. Cameron and J. H. van Lint. *Graphs, Codes and Designs*. LMS Lecture Note Series 43, Cambridge University Press, Cambridge etc., 1980.
9. J. W. S. Cassels. *Rational Quadratic Forms*. Academic Press, London New York San Francisco, 1978.
10. Y. Choie and E. Jeong. Jacobi forms over totally real fields and codes over $\mathbb{F}_p$. *Illinois J. Math.*, 46:627–643, 2002.
11. J. H. Conway. A characterisation of Leech's lattice. *Invent Math.*, 7:137–142, 1969.
12. J. H. Conway. Three lectures on exceptional groups. In M. B. Powell and G. Higman, editors, *Finite Simple Groups*, pages 215–247, New York, 1971. Academic Press.
13. J. H. Conway. The automorphism group of the 26-dimensional even Lorentzian lattice. *J. Algebra*, 80:159–163, 1983.
14. H. S. M. Coxeter. Extreme forms. *Canad. J. Math.*, 3:391–441, 1951.
15. J. H. Conway and S. P. Norton. Monstrous moonshine. *Bull. Lond. Math. Soc.*, 11:308–339, 1979.
16. J. H. Conway and V. Pless. On the enumeration of self-dual codes. *J. Combin. Theory, Ser. A*, 28:26–53, 1980.
17. J. H. Conway, R. A. Parker, and N. J. A. Sloane. The covering radius of the Leech lattice. *Proc. R. Soc. Lond.*, A 380:261–290, 1982.
18. J. H. Conway and N. J. A. Sloane. Leech roots and Vinberg groups. *Proc. R. Soc. Lond.*, A 384:233–258, 1982.
19. J. H. Conway and N. J. A. Sloane. Lorentzian forms for the Leech lattice. *Bull. Am. Math. Soc.*, 6:215–217, 1982.
20. J. H. Conway and N. J. A. Sloane. Twenty-three constructions for the Leech lattice. *Proc. R. Soc. Lond.*, A 381:275–283, 1982.
21. J. H. Conway and N. J. A. Sloane. *Sphere Packings, Lattices and Groups*. Springer, New York etc., 1988.

22. W. Ebeling. Strange duality, mirror symmetry, and the Leech lattice. In J. W. Bruce, D. Mond, editors, *Singularity Theory (London Math. Soc. Lecture Note Ser., Vol.* 263*)*, pages 55–77, Cambridge University Press, Cambridge, 1999. (Proceedings of the European Singularities Conference, Liverpool, 1996)
23. W. Ebeling. Lattices and Codes. In St. Löwe, F. Mazzocca, N. Melone, U. Ott, editors, *Methods of discrete mathematics (Quad. Mat.,* 5*)*, pages 103–143, Aracne, Rome, 1999. (Summer School, Braunschweig, 1999)
24. M. Eichler. On theta functions of real algebraic number fields. *Acta Arithmetica*, 33:269–292, 1977.
25. E. Freitag. *Hilbert Modular Forms*. Springer, Berlin etc., 1990.
26. G. van der Geer. *Hilbert Modular Surfaces*. Springer, Berlin etc., 1988.
27. M. J. E. Golay. Notes on digital coding. *Proc. IRE (IEEE)*, 37:657, 1949.
28. R. C. Gunning. *Lectures on Modular Forms*. Annals of Math. Studies. Princeton University Press, Princeton, 1962.
29. R. W. Hamming. A theory of self-checking and self-correcting codes – case 20878, memorandum 48-110-31. Technical report, Bell Telephone Laboratories, June 1948.
30. R. W. Hamming. Error detecting and error correcting codes. *Bell System Tech. J.*, 29:147–160, 1950.
31. D. Hilbert and S. Cohn-Vossen. *Anschauliche Geometrie*. Julius Springer, Berlin, 1932. Engl. translation *Geometry and the Imagination*, Chelsea, New York, 1952.
32. E. Hecke. Analytische Arithmetik der positiven quadratischen Formen. *Kgl. Danske Videnskabernes Selskab. Math.-fys. Medd.*, 17(12), 1940. (*Mathematische Werke*, pages 789–918, Vandenhoeck und Ruprecht, Göttingen, 1959).
33. F. Hirzebruch. Hilbert modular surfaces. *L'Enseignement Math.*, 19:183–282, 1973. (also Monographie No. 21 de L'Enseignement Math., Genève, 1973).
34. F. Hirzebruch. The ring of Hilbert modular forms for real quadratic fields of small discriminant. In *Modular Functions of One Variable VI, Proceedings, Bonn 1976 (Lect. Notes in Math. No. 627)*, pages 288–323, New York etc., 1977. Springer. (*Gesammelte Abhandlungen*, Vol. II, pages 501–536, Springer, Berlin etc., 1987).
35. F. Hirzebruch. The icosahedron. Raymond and Beverly Sackler Distinguished Lectures in Mathematics, Tel Aviv University, Tel Aviv (*Gesammelte Abhandlungen*, Vol. II, pages 656–661, Springer, Berlin etc., 1987), March 1981.
36. F. Hirzebruch. *Gesammelte Abhandlungen. Collected Papers*. Springer, Berlin etc., 1987.
37. F. Hirzebruch. Codierungstheorie und ihre Beziehung zu Geometrie und Zahlentheorie. In *Vorträge Rheinisch-Westfälische Akademie der Wissenschaften, N 370*, Opladen, 1989. Westdeutscher Verlag.
38. F. Hirzebruch. *Manifolds and Modular Forms*. Friedrich Vieweg und Sohn, Braunschweig Wiesbaden, 1992.
39. J. E. Humphreys. *Introduction to Lie Algebras and Representation Theory*. Springer, New York etc., 1972.
40. F. Klein. Weitere Untersuchungen über das Ikosaeder. *Math. Ann.*, 12, 1877. (*Gesammelte mathematische Abhandlungen* Bd. II, pages 321–380, Springer, Berlin, 1922 (Reprint 1973)).
41. F. Klein. *Lectures on the Icosahedron and the Solution of Equations of the Fifth Degree*. Dover, New York, 1956.
42. F. Klein. *Vorlesungen über das Ikosaeder und die Auflösung der Gleichungen vom fünften Grade. Herausgegeben mit einer Einführung und mit Kommentaren von Peter Slodowy*. Birkhäuser, Basel Boston Berlin, 1993.
43. H. D. Kloosterman. Thetareihen in total reellen algebraischen Zahlkörpern. *Math. Ann.*, 103:279–299, 1930.
44. M. Kneser. Klassenzahlen definiter quadratischer Formen. *Archiv Math.*, 8:241–250, 1957.
45. M. Kneser. *Quadratische Formen*. Revised and edited in collaboration with Rudolf Scharlau. Springer-Verlag, Berlin, 2002.
46. H. V. Koch. Unimodular lattices and self-dual codes. In *Proc. International Congress of Mathematicians, Berkeley*, pages 457–465, 1986.
47. H. Koch. The completeness principle for the Golay codes and some related codes. In M. M. Arslanov, A. N. Parshin, I. R. Shafarevich, editors, *Algebra and Analysis*, pages 75–80, Walter de Gruyter & Co., Berlin New York, 1996. (Intern. Conference, Kazan, 1994)

48. H. Koch and B. B. Venkov. Über gerade unimodulare Gitter der Dimension 32, III. *Math. Nachr.*, 152:191–213, 1991.
49. A. Korkine and G. Zolotareff. Sur les formes quadratiques positives quaternaires. *Math. Ann.*, 5:581 583, 1872.
50. A. Korkine and G. Zolotareff. Sur les formes quadratiques. *Math. Ann.*, 6:366–389, 1873.
51. A. Korkine and G. Zolotareff. Sur les formes quadratiques positives. *Math. Ann.*, 11:242–292, 1877.
52. K. Lamotke. *Regular Solids and Isolated Singularities*. Friedrich Vieweg und Sohn, Braunschweig Wiesbaden, 1986.
53. S. Lang. *Introduction to Modular Forms*. Springer, Berlin etc., 1976.
54. J. Leech. Notes on sphere packings. *Canad. J. Math.*, 19:251–267, 1967.
55. J. H. van Lint. *Coding Theory*. Lecture Notes in Math. 201. Springer, Berlin etc., 1971.
56. J. H. van Lint. *Introduction to Coding Theory*. Springer, New York etc., 1982.
57. J. H. van Lint and G. van der Geer. *Introduction to Coding Theory and Algebraic Geometry*. DMV Seminar, Band 12. Birkhäuser, Basel Boston Berlin, 1988.
58. J. S. Leon, V. Pless, and N. J. A. Sloane. Self-dual codes over $GF(5)$. *J. Combin. Theory Ser. A*, 32:178–194, 1982.
59. E. Looijenga and Ch. Peters. Torelli theorems for Kähler K3 surfaces. *Compositio Math.*, 42:145–186, 1981.
60. J. Leech and N. J. A. Sloane. Sphere packings and error-correcting codes. *Canad. J. Math.*, 23:718–745, 1971.
61. J. Milnor and D. Husemoller. *Symmetric Bilinear Forms*. Springer, Berlin Heidelberg New York, 1973.
62. J. Milnor. On simply connected 4-manifolds. In *Proc. Symposium Internacional Topologia Algebraica, Mexico*, pages 122–128, 1958.
63. H. Minkowski. Grundlagen für eine Theorie der quadratischen Formen mit ganzzahligen Koeffizienten, 1884. (*Gesammelte Abhandlungen* Bd. I, pages 3–144, Teubner, Leipzig, 1911 (Reprint Chelsea, New York, 1967)).
64. F. J. MacWilliams, C. L. Mallows, and N. J. A. Sloane. Generalization of Gleason's theorem on weight enumerators of self-dual codes. *IEEE Trans. Inform. Theory*, IT 18:794–805, 1972.
65. L. J. Mordell. *Diophantine Equations*. Academic Press, New York, 1969.
66. F. J. MacWilliams and N. J. A. Sloane. *The Theory of Error-Correcting Codes*. North-Holland, Amsterdam etc., 1977.
67. G. Nebe. An even unimodular 72-dimensional lattice of minimum 8. Preprint, arXiv: 1008.2862, 2010.
68. H.-V. Niemeier. Definite quadratische Formen der Dimension 24 und Diskriminante 1. *J. Number Theory*, 5:142–178, 1973.
69. V. V. Nikulin. Integral symmetric bilinear forms and some of their applications. *Izv. Akad. Nauk. SSSR Ser. Mat.*, 43:111–177, 1979. Engl. translation in *Math. USSR Izv.*, 14, No. 1, 103–167, 1980.
70. A. Neumaier and J. J. Seidel. Discrete hyperbolic geometry. *Combinatorica*, 3:219–237, 1983.
71. A. Ogg. *Modular Forms and Dirichlet Series*. Benjamin, New York, 1969.
72. O. T. O'Meara. *Introduction to Quadratic Forms*. Springer, Berlin Heidelberg New York, third corrected printing (1973) edition, 1963.
73. A. M. Odlyzko and N. J. A. Sloane. New bounds on the number of unit spheres that can touch a unit sphere in $n$ dimensions. *J. Combin. Theory Ser. A*, 26:210–214, 1979.
74. U. Ott. Local weight enumerators for binary self-dual codes. *J. Combin. Theory Ser. A*, 86:362–381, 1999.
75. V. Remmert. *Über Codes, Gruppen und Gitter*. Diplomarbeit, Freiburg, 1992.
76. P. Samuel. *Théorie Algébrique des Nombres*. Hermann, Paris, 1967. Engl. translation *Algebraic Number Theory*, Houghton Mifflin, New York, 1975.
77. K. Saito. Duality for regular systems of weights: a précis. In M. Kashiwara, A. Matsuo, K. Saito, I. Satake, editors, *Topological Field Theory, Primitive Forms and Related Topics (Progress in Math., Vol. 160)*, pages 379–426, Birkhäuser, Boston Basel Berlin, 1998.
78. B. Schoeneberg. Das Verhalten von mehrfachen Thetareihen bei Modulsubstitutionen. *Math. Ann.*, 116:511–523, 1939.

79. B. Schoeneberg. *Elliptic Modular Functions*. Springer, Berlin etc., 1974.
80. R. L. E. Schwarzenberger. *N-Dimensional Crystallography*. Pitman, San Francisco, 1980.
81. J.-P. Serre. *A Course in Arithmetic*. Springer, New York etc., 1973.
82. J.-P. Serre. *Complex Semisimple Lie Algebras*. Springer, New York etc., 1987.
83. N.-P. Skoruppa. Thetareihen über total reellen Zahlkörpern. Unpublished manuscript, Max-Planck-Institut für Mathematik, Bonn, 1987.
84. N. J. A. Sloane. Binary codes, lattices and sphere packings. In P. J. Cameron, editor, *Combinatorial Surveys*, pages 117–164, New York, 1977. Academic Press.
85. N. J. A. Sloane. Self-dual codes and lattices. In *Proc. Symp. Pure Math. Vol. 34*, pages 273–308, 1979.
86. H. J. S. Smith. On the orders and genera of quadratic forms containing more than three indeterminates. *Proc. R. Soc. Lond.*, pages 197–208, 1867. (*Coll. Math. Papers* I, 510–523).
87. I. N. Stewart and D. O. Tall. *Algebraic Number Theory*. Chapman and Hall, London New York, second edition, 1987. First edition published in 1979.
88. Th. M. Thompson. *From Error-Correcting Codes through Sphere Packings to Simple Groups*. Math. Assoc. Am., Washington, DC, 1983.
89. B. B. Venkov. On the classification of integral even unimodular 24-dimensional quadratic forms. *Trudy Mat. Inst. Steklov*, 148:65–76, 1978. Engl. translation in *Proc. Steklov Inst. Math.*, 4:63–74, 1980.
90. È. B. Vinberg. Some arithmetical discrete groups in Lobačevskiĭ spaces. In *Discrete Subgroups of Lie Groups and Applications to Moduli*, pages 323–348, Oxford, 1975. Oxford University Press. (Intern. Colloq., Bombay, 1973).
91. Z.-X. Wan. On the uniqueness of the Leech lattice. *Europ. J. Combinatorics*, 18:455–459, 1997.
92. H. Weber. *Lehrbuch der Algebra*, volume II. Friedrich Vieweg und Sohn, Braunschweig, second edition, 1899.
93. E. Witt. Über Steinersche Systeme. *Abh. Math. Sem. Univ. Hamburg*, 12:265–275, 1938.
94. E. Witt. Spiegelungsgruppen und Aufzählung halbeinfacher Liescher Ringe. *Abh. Math. Sem. Univ. Hamburg*, 14:289–322, 1941.

# Index